Light-Emitting Diodes - New Perspectives

Edited by Chandra Shakher Pathak and Uday Dadwal

Published in London, United Kingdom

Light-Emitting Diodes - New Perspectives
http://dx.doi.org/10.5772/intechopen.100741
Edited by Chandra Shakher Pathak and Uday Dadwal

Contributors
Hao-Chung Kuo, Tingwei Lu, Tingzhu Wu, Yen-Wei Yeh, Zhong Chen, Abdul Shabir, Cher Ming Tan, Rajamouli Boddula, Jyoti Singh, Niteen P. Borane, Li Wang, Hideki Hirayama, Thamer A. Tabbakh, Deepak Anandan, Michael J. Sheldon, Prashant Tyagi, Ahmad Alfaifi

First published in London, United Kingdom, 2023 by IntechOpen
IntechOpen is the global imprint of INTECHOPEN LIMITED, registered in England and Wales, registration number: 11086078, 5 Princes Gate Court, London, SW7 2QJ, United Kingdom

British Library Cataloguing-in-Publication Data
A catalogue record for this book is available from the British Library

Additional hard and PDF copies can be obtained from orders@intechopen.com

Light-Emitting Diodes - New Perspectives
Edited by Chandra Shakher Pathak and Uday Dadwal
p. cm.
Print ISBN 978-1-80356-560-6
Online ISBN 978-1-80356-561-3
eBook (PDF) ISBN 978-1-80356-562-0

Meet the editors

Dr. Chandra Shakher Pathak received his Ph.D. in Physics from the Indian Institute of Technology, Delhi, India. He obtained an MSc in Physics from Kumaun University Nainital, Uttarakhand, India, and an MTech (Gold medalist) in Advanced Materials Science and Technology from the National Institute of Technology, Durgapur, India. His research work focuses on nanomaterials, thin film growth, characterization, photovoltaics, organic–inorganic heterojunctions, Schottky diodes, and 2D materials. Currently, Dr. Pathak is researching the fabrication and stability of perovskite solar cells.

Dr. Uday Dadwal is currently a technical officer at the Nanoscale Research Facility (NRF), Indian Institute of Technology Delhi (IIT Delhi), New Delhi, India. He obtained an MSc in Physics from Jiwaji University, Gwalior, India, in 2005. He obtained a Ph.D. from the Centre for Applied Research in Electronics, IIT Delhi in 2013 with work related to the investigation of blistering and exfoliation phenomena in hydrogen-implanted semiconductors. After obtaining his Ph.D., Dr. Dadwal became a project scientist at the NRF. He was involved in the setting up of state-of-the-art nanofabrication facilities in the class-100 and class-1000 clean rooms such as electron beam lithography, maskless photolithography, atomic layer deposition, reactive ion etching, oxidation furnace, and chemical vapor deposition. He was also involved in the growth of semiconductor nanostructures for water-splitting applications. His research has focused on layer-splitting studies of wide-bandgap semiconductors including GaN and AlN, wafer bonding studies using hydrogen ion implantation, and growth and study of semiconductor nanostructures for water-splitting applications.

Contents

Preface **IX**

Section 1
Micro and Nano LEDs **1**

Chapter 1 **3**
Full-Color Micro-LED Devices Based on Quantum Dots
by Tingzhu Wu, Tingwei Lu, Yen-Wei Yeh, Zhong Chen and Hao-Chung Kuo

Chapter 2 **25**
Milestone Developments and New Perspectives of Nano/Nanocrystal Light Emitting Diodes
by Jyoti Singh, Niteen P. Borane and Rajamouli Boddula

Section 2
High Power LEDs **47**

Chapter 3 **49**
High-Temperature Operating Narrow-Period Terahertz Quantum Cascade Laser Designs
by Li Wang and Hideki Hirayama

Chapter 4 **69**
Degradation Analysis of Silicone as Encapsulation and Molding Material in High Power LEDs
by Abdul Shabir and Cher Ming Tan

Section 3
GaN LEDs **89**

Chapter 5 **91**
Recent Advancements in GaN LED Technology
by Thamer A. Tabbakh, Deepak Anandan, Michael J. Sheldon, Prashant Tyagi and Ahmad Alfaifi

Preface

This book provides a comprehensive and current overview of light-emitting diodes (LEDs). It discusses recent advances in micro, nano, and high-power LEDs and reviews the various degradation analyses of LEDs and gallium nitride (GaN)-based LEDs. The latter were discovered by Shuji Nakamura, who won the 2014 Nobel Prize in Physics along with Isamu Akasaki and Hiroshi Amano. After the invention of GaN-based blue LEDs, there was a revolutionary development in the III-nitride-based semiconductor industry worldwide.

This book is divided into three sections. The first section presents an introduction to micro-LEDs based on quantum dots and developments in nano LEDs. This section includes the chapters "Full-Color Micro-LED Devices Based on Quantum Dots" and "Milestone Developments and New Perspectives of Nano/Nanocrystal Light Emitting Diodes".

The second section discusses high-power LEDs. It includes the chapters "High-Temperature Operating Narrow-Period Terahertz Quantum Cascade Laser Designs" and "Degradation Analysis of Silicone as Encapsulation and Molding Material in High Power LEDs".

The third section includes one chapter on "Recent Advancements in GaN LED Technology".

We thank all the authors for their contributions and patience in editing the book.

We also thank Author Service Manager Ana Cink and IntechOpen for allowing us to edit this fascinating book project.

Chandra Shakher Pathak
Ben-Gurion University of the Negev,
Beersheba, Israel

Uday Dadwal
Indian Institute of Technology Delhi,
New Delhi, India

Section 1

Micro and Nano LEDs

Chapter 1

Full-Color Micro-LED Devices Based on Quantum Dots

Tingzhu Wu, Tingwei Lu, Yen-Wei Yeh, Zhong Chen and Hao-Chung Kuo

Abstract

Quantum dots (QDs) show remarkable optical and electrical characteristics. They offer the advantage of combining micro-LEDs (μLEDs) for full-color display devices due to their exceptional features. In addition, μLED used in conjunction with QDs as color-conversion layers also provide efficient white LEDs for high-speed visible light communication (VLC). In this article, we comprehensively review recent progress in QD-based μLED devices. It includes the research status of various QDs and white LEDs based on QDs' color conversion layers. The fabrication of QD-based high-resolution full-color μLEDs is also discussed. Including charge-assisted layer-by-layer (LbL), aerosol jet printing, and super inkjet printing methods to fabricate QD-based μLEDs. The use of quantum dot photoresist in combination with semipolar μLEDs is also described. Finally, we discuss the research of QD-based μLEDs for visible light communication.

Keywords: quantum dots, micro-LEDs, full-color displays, visible light communication

1. Introduction

Light-emitting diodes (LED) have been widely used in daily life due to their advantages of high efficiency, energy saving, and long working life, such as lighting sources, full-color displays, and backlight sources of liquid crystal displays [1]. Recently, mobile device displays, such as smart furniture, augmented reality (AR), virtual reality (VR), and wearable devices have piqued extensive attention from the semiconductor industries and research on micro-LEDs (μLEDs), which are referred to as chips with sizes less than 50 × 50 μm^2 [2]. Due to their excellent properties in terms of brightness, lifetime, resolution, and efficiency, μLEDs have been considered the most promising next-generation display technologies [3]. The potential of μLEDs to replace conventional display technologies is owing to the combination of their self-emissive mechanism and inorganic material characteristics. Over the past decade, the number of commercially available μLED displays has grown significantly, as manufacturers seek to capitalize on the success of this technology. Sony introduced its first 55-inch full high-definition (HD) μLED TV panel with 1920 × 1080 resolution in 2012, which consists of over 6 million individual μLEDs. Samsung unveiled the world's first consumer modular μLED 146-inch TV in 2018, which is named "The

Wall." In 2022, Samsung will launch a new μLED TV with 25 million pixels, providing vivid colors,high definition, and contrast, and also supports 20-bit grayscale depth, more than 1 million levels of brightness and color. In the academic field, μLEDs have been studied for more than 10 years.

In addition to displays, μLEDs have recently been adopted as transmitters in visible light communication (VLC) systems, based on their quick response times [4]. Currently, radio frequency (RF) communication faces some challenges, such as interference, bandwidth limitations, security issues, transmission power limitations, and power inefficiency [5]. VLC is an emerging technology that addresses the crowded radio spectrum, using visible light to communicate to enable high-speed internet connections. As a light source that is harmless to human body, LED can have both lighting and communication functions when used in VLC, which can save extra power [6].

The modulation bandwidth of an LED is constrained by the carrier lifetime and the time constant consisting of capacitance of a depletion layer and junction differential resistance. Due to their small size, μLEDs can withstand higher injection current densities, thereby enabling smaller carrier lifetimes and higher modulation bandwidths. In addition, a smaller active area will reduce the geometric capacitance of the device, thereby reducing the RC time constant. In addition, μLEDs also have better current uniformity, which will also increase the modulation bandwidth featuring exclusive properties making μLEDs widely applicable for high-speed VLC system.

2. Full-color μLED display based on quantum dots

2.1 Background of full-color μLED display

The commonly used full-color strategy is employing the combination of red–green–blue (RGB) μLED devices in a display. However, this approach has a number of drawbacks. First, the so-called "green gap" created by green μLEDs results in low efficiency [7]. For green LEDs, a high proportion of indium is required in the active region, which requires relatively low growth temperatures, resulting in poor crystal quality of the LED epitaxial layers. In addition, high proportions of indium produce strong polarization fields in InGaN/GaN multiple quantum wells (MQWs) and lead to strong quantum-confined Stark effects, reducing recombination efficiency [7]. Also, red μLEDs are problematic. The active region of the red LED consists of AlGaInP material, which has a high surface recombination velocity (~10^6 cm/s) [8], coupled with a long carrier diffusion length of about a few microns, making non-radiative surface recombination much more efficient [9]. Therefore, as the device size shrinks to a few microns, the EQE degradation of red μLEDs is more severe than that of blue and green μLEDs. Another problem in the RGB μLED strategy is the mismatch of drive voltages between RGB pixels. The threshold voltage of the blue LED is about 3.3 V, while the threshold voltages of the red and green LEDs are 1.7 and 2.2 V, which complicates the driver circuit design.

To address these issues, blue μLEDs can be integrated with color converters, such as yellow-emitting phosphors or red and green-emitting nanocrystals (NCs), for higher-quality full-color displays [10]. To date, extensive research and development have been carried out on phosphor materials for PC-LEDs. Lin. Wait. Successfully fabricated high luminance efficiency and wide color gamut for NC-based solid-state and hybrid WLED devices, which exhibited higher efficiency (51 lm W^{-1}), wide color gamut (122% of NTSC and 91% of Rec. 2020), and the efficiency decays by about 12% during the 200-h reliability test [11]. However, organic or inorganic phosphors are generally not suitable for μLED displays due to their spectral width and asymmetry, inherent instability, low red phosphor efficiency, down conversion energy loss, and low absorption cross-section in the blue/UV wavelength region [12]. In addition, the particle size of conventional phosphors may be comparable to or larger than that of μLED chips, which will affect device performance.

2.2 QD-based color conversion for μLEDs

Quantum dots (QDs) are nanoscale semiconductor crystals, and whose electrical and optical properties can be tuned by changing their sizes [13, 14]. Compared with traditional fluorophores, several QD photophysical properties are distinct and unmatched. The first is the ability to tune photoluminescence emission according to the core size and quantum confinement effects of semiconductor binary combinations. This unique advantage allows one to control the emission properties of QDs by controlling the core size [1]. In addition, QDs have broad absorption spectra, starting with blue emission and then steadily increasing towards UV. As combined with the aforementioned properties, quantum dots have emerged as omnidirectional attractive fluorophores for full-color displays.

Core-type QDs, such as CdSe, CdS, or CdTe, are the most studied and commercialized QDs due to their excellent optical and electrical properties [15, 16]. Using these precursors, high fluorescence quantum yields of up to 80% have been reported [17, 18]. However, many heavy metal elements are contained in such quantum dots, and Core-type QDs will pollute the environment, which limits their wide application. To prevent the use of toxic metal elements for the persistent development of QD-based products, the fabrication of Cd/Pb-free QDs has become a subject of intense interest in recent years. In 2020, Soheyli et al. reported a novel aqueous-phase approach for the preparation of multicomponent In-based QDs. Absorption and photoluminescence emission spectra of the as-prepared QDs were tuned by alteration of QDs' composition as Zn-Ag-In -S/ZnS, Ag-In-S/ZnS, and Cu-Ag-In-S/ZnS core/shell QDs [19]. However, such materials have extremely small color gamut and color purity and are not suitable for display applications [20].

As an environmentally benign material belonging to the III–V group semiconductor, InP is deemed as another promising candidate to replace Cd/Pb-based QDs [21]. InP QDs are typical III–V group semiconductor nanocrystals that feature large excitonic Bohr radius and high carrier mobility [22]. Owing to the merits of low toxicity, high QYs, and broad color tunability, InP QDs are particularly suitable to construct LEDs for indoor/outdoor illumination, traffic signaling, and liquid-crystal display backlighting [23]. In general, InP QDs can be applied in μLEDs either as a photoluminescent layer or an electroluminescent layer. In 2015, Zhang et al. demonstrated a single "cadmium-free" component system consisting of Cu-doped InP core/ZnS barrier/InP quantum well/ZnS shell QDs [24]. These QDs exhibit two emission peaks by controlling the barrier thickness under single-excitation wavelength, one of which is attributed to Cu-doped InP, and the other resulting from InP quantum well. Using optimal structures as color converters, the WLED was obtained with a color rendering index (CRI) up to 91 and CIE color coordinates of (0.338, 0.330) by combination with blue LED chip.

The metal halide perovskites QDs (PQDs) are considered the next generation thin-film LED light emitters because of their outstanding characteristics compared to colloidal QDs, which are derived from their narrower FWHM (below 30 nm) [25]. Tunable narrow and symmetrical PL peaks in the visible spectral range can be achieved by tuning the halide composition in perovskite materials. This property yields greater color purity, higher photoluminescence quantum yield (approaching 100%), more convenient synthesis, and lower manufacturing costs than conventional QDs [26]. In addition, PQDs have short carrier recombination lifetimes, which also lays the foundation for their application in the development of white-light VLC systems with high modulation bandwidth [27]. The ability to easily handle colloidal PQDs in solution enables the production of cost-effective large-area light-emitting layers [28]. Therefore, PQDs have been extensively studied in electronic and optoelectronic applications, such as photodetection, photovoltaics, and photoemission; especially as active materials or color converters in LEDs. To date, various PQD-based white LEDs have been reported using diverse structural designs that feature different advantages and disadvantages. However, due to poor thermal resistance and instability under high energy radiation, most PQD-based white light-emitting diodes (WLEDs) show only modest luminous efficiency of approximately 50 lm/W and a short lifetime of <100 h.

In 2019, Kang et al. demonstrated a new type of PQD film called PQD paper by using cellulose nanocrystals (CNCs) [29]. **Figure 1(a)** schematically shows the fabrication process of the PQD paper. The CNC suspension and $CH_3NH_3PbBr_3$ are combined in dimethylformamide, upon which the mixture is dried on the membrane to produce the PQDs paper. As shown in **Figure 1(b)**, the entangled CNC structure was clearly demonstrated by SEM. With the help of the CNC structure, the PQD paper performs the flexibility and unique mechanical strength. The size of the PQDs was $\approx 3^{-8}$ nm, which can enhance the light emission of perovskite by the provided strong quantum confinement, as shown in **Figure 1(c)**. **Figure 1(d)** illustrates the flux and current-dependent luminous efficiency of the PQD paper-based LED, the inset shows the emission of white LED. It can be clearly observed that the PQD paper-based LED has an ultrahigh luminous efficiency of 124 lm W^{-1} and still exhibits a luminous efficiency of over 100 lm W^{-1} even when the drive current increases to 50 mA. Moreover, after continuously working for 240 h, the PQD paper-based LED can maintain 87.6% luminous flux. Compared to the normal flat design, by using curved PQD paper, the viewing angle of LED was increased from 120° to 143° benefiting from the flexible nature of paper. This work also shows the fabricated white LEDs have a wide color gamut of 123% of NTSC standard and 92% of Rec. 2020, as shown in **Figure 1(e)**.

2.3 Fabrication of QD-based μLEDs for high-resolution displays

QD-based display technology has proven to be ideal for next-generation displays. In order to meet the needs of achieving high-resolution display, RGB QDs with sufficiently high pixel density need to be deposited on top of the substrate surface of μLED array. The μLED device realized in this way can replace the traditional OLED or LCD display to realize an ultra-high-definition display, for example, to meet the needs of AR/VR or ultra-high-definition TV. QD deposition and patterning techniques on a selected area of a substrate remain the major bottlenecks in realizing such devices. Therefore, in recent years, researchers have been exploring facile and efficient QDs patterning techniques, such as spray coating, aerosol jet printing, super-inkjet printing, etc.

To accurately locate QDs on the μLED array, the inkjet printing (IJP) has become a key technology for realizing QD-LED display. In 2015, Han et al. combined ultraviolet (UV) μLED and colloidal QDs to achieve a full-color display. RGB QDs were deposited on an array of GaN-based UV μLEDs with the help of an aerosol inkjet

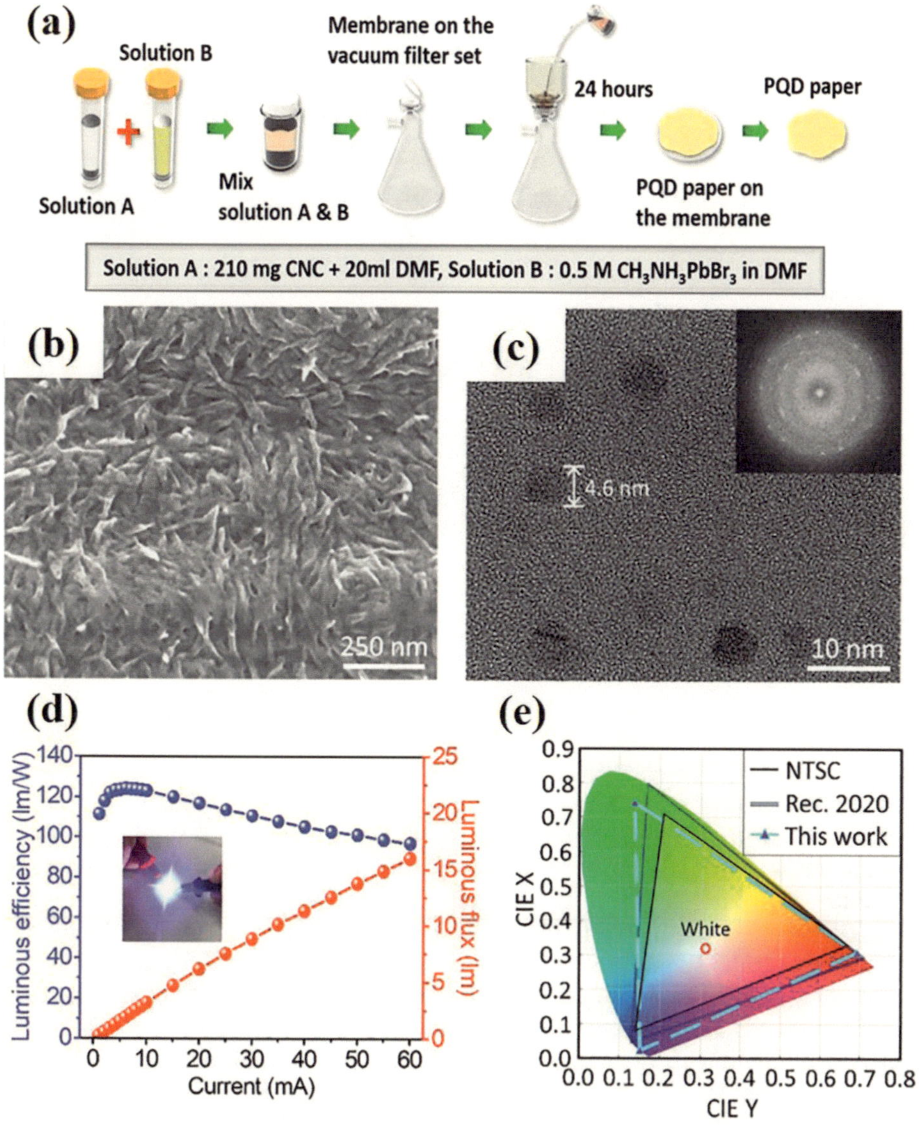

Figure 1.
(a) Schematic of the fabrication process of the PQD paper. (b) SEM image of the PQD paper surface. (c) TEM image of the $CH_3NH_3PbBr_3$ PQDs obtained from the paper. The electron diffraction pattern in the inset reveals the high crystallinity of the PQDs. (d) Current-dependent luminous efficiency and luminous flux of the PQD paper-based LED. The inset shows the emission of white LED. (e) CIE diagram illustrating the color gamut of the NTSC standard, the Rec. 2020 standard, and the PQD paper-based LED [29]. Figure reproduced with permission from John Wiley and Sons.

printer to ensure fine printing that is highly precise and mask-less and to enable noncontact deposition of liquids containing functional materials [30]. MQWs μLED arrays are grown on sapphire substrates. The UV micro-LED array is fabricated on a UV epitaxial wafer with a peak wavelength of 395 nm. Since pixels in the same column share an electrode of n-type GaN, the GaN is dry-etched into the sapphire substrate to create isolation trenches to isolate all micro-LED arrays. When the array is complete, the RGB QDs are sequentially sprayed onto the microLED array using

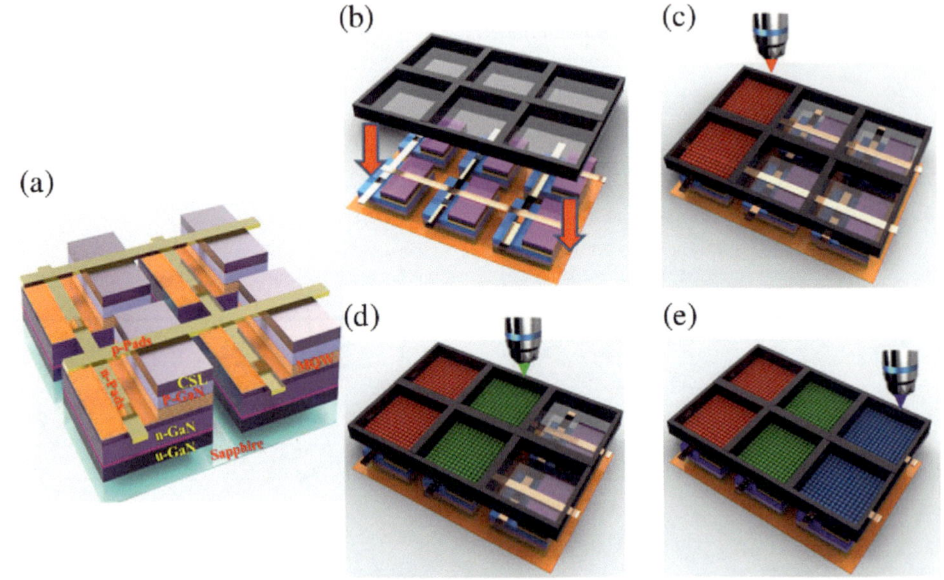

Figure 2.
Process flow of the full-color microdisplay. (a) The structure of the micro-LED arrays. (b) Aligning the mold to the UV micro-LED array. (c)–(e) Consequently jetting the RGB QDs inside the molded window to form the full-color pixels [31]. Figures reproduced with permission from Optica Publishing Group.

aerosol jet printing. The concentration of RGB QDs is about 5 mg/ml. Among the process parameters of aerosol jet printing, the working distance, table speed, carrier gas flow, sheath gas flow, and atomization frequency need to be adjusted to obtain a spraying line width of 35 μm. The working distance and table speed between the nozzle and the substrate were 1 mm and 10 mm/s, respectively. However, the cross-talk effect still occurred during the QD deposition due to the overflow of QDs during the solvent evaporation. To address the problem of cross-talk during QD deposition, the position of the QD must be restricted. In 2017, Lin et al. demonstrated a significant reduction in the cross-talk effect during the AJ printing process by using a photoresist (PR)-defined mold with a blocking wall to confine QDs [31]. **Figure 2** shows the fabrication process of the monolithic device. First, UV micro-LED arrays were fabricated on UV wafers with a peak wavelength of 395 nm and a pitch size of 40 μm. Isolation trenches are formed by etching GaN on a sapphire substrate. Through the dry etching process, SiO_2 is used as a hard mask. Finally, p-electrode stripes are defined on top of the chip, n-electrode stripes are defined on the n-GaN layer, and all pixels in the same row are connected. By aligning the window of the mold with the micro-LED mesa, as shown in **Figure 2(b)–(e)**, AJ RGB QDs can be efficiently deposited on the micro-LED mesa area without overlapping the trench area. The printing parameters of different quantum dots need to be optimized. For blue QDs, the working distance is 1 mm, the carrier gas flow rate is 66 sccm, the sheath gas flow rate is 11 sccm, and the stage speed is 10 mm/s. For green QDs, the working distance is 1 mm, the carrier gas flow rate is 72 sccm, the sheath gas flow rate is 15 sccm, and the stage speed is 10 mm/s. For red QDs, the working distance is 1 mm, the carrier gas flow rate is 83 sccm, the sheath gas flow rate is 17 sccm, and the stage speed is 10 mm/s.

Pixels of an RGB display are demonstrated after depositing QDs on micro-LEDs using AJ on a die that has been window-confined. **Figure 3(a)** shows the microscope

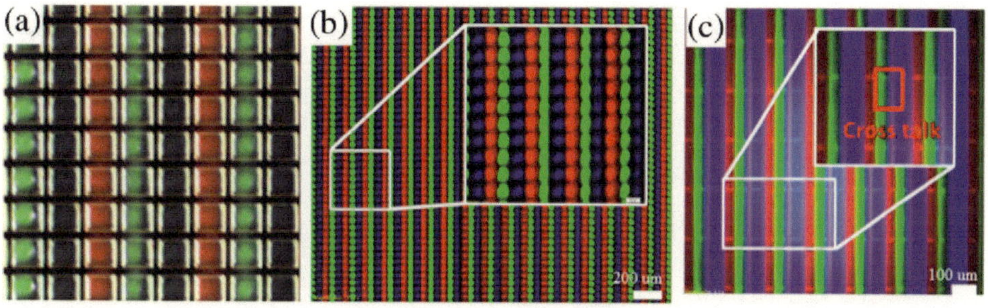

Figure 3.
(a) Microscope image of the full-color micro-LED after jetted QDs in the PR mold. (b) The RGB pixel array observed by fluorescence microscopy. (c) The fluorescence microscopy image of the jetted QD pixels without the PR mold [31]. Figures reproduced with permission from Optica Publishing Group.

image of the full-color micro-LED after jetted QDs in the PR mold. **Figure 3(b)** shows the RGB pixel array observed by fluorescence microscopy, indicating that the luminescent regions of quantum dots have clear boundaries. Compared with the result of printing without PR window restriction as shown in **Figure 3(c)**, the fluorescence image shows no obvious separation between the printed QDs. Clearly, crosstalk is observed in the blue QD lines.

In 2019, Chen et al. proposed hybrid QD nanoring μLEDs (QD-NR-μLEDs) fabricated by QD printing and electron beam (E-beam) lithography [32]. There are three parts to this device, namely, a normal green LED, a blue NR-μLED, and a red QD-NR-μLED, and each region can be regarded as a subpixel, as shown in the SEM images of **Figure 4(a)**. Besides, the nonradiative resonant energy transfer (NRET) mechanism was allowed for adjacent coupling between the exposed InGaN/GaN MQW sidewalls and QDs. During the manufacturing process, an electron beam lithography system is first used to define an area with negative photoresist, which is used as the RGB pixel. The green pixels are all shaped into rectangular mesas. The remaining red and blue regions form NR arrays with hexagonal close packing. After that, nickel is deposited by electron beam evaporation, and then the photoresist is removed by a lift-off process to form a hard mask pattern. Subsequently, by using Inductively Coupled Plasma Reactive Ion Etching (ICP-RIE), the GaN-based material is etched to define the active regions, separating the pn layers and isolating each subpixel. Next, residual nickel is removed by HCl solution. The ALD technology was used to deposit the Al_2O_3 passivation layer. In order to create color conversion layer, CdSe/ZnS red QDs were sprayed on a region of blue NR-μLED via the super-inkjet (SIJ) printing system. Then, spin-on glass (SOG) was used to isolate the pn electrodes and protect the QD layer. After, a transparent conducting oxide (TCO) layer was orderly deposited via the hole process by SOG etching and Ni/Au metal deposition for the pn electrodes with the lift-off process. Finally, a distributed Bragg reflector (DBR) was used to recycle blue light and cover the red color region to filter out. **Figure 4(b)** showed the SEM image of NR-μLED with 30° tilt angle. Besides, it can be clearly observed from the transmission electron microscopy (TEM) images in **Figure 4(c)** that the sidewalls of MQWs were closely surrounded by QDs, which is important to the NRET mechanisms.

The overlapping relationship between the absorption spectrum of red QDs and the EL spectrum of a blue NR-μLED was demonstrated in **Figure 5(a)**. The emission wavelength of the blue NR-μLED is 467 nm, which is in the range of intense QD absorption. That indicates good spectral overlaps of MQWs with QD absorption and

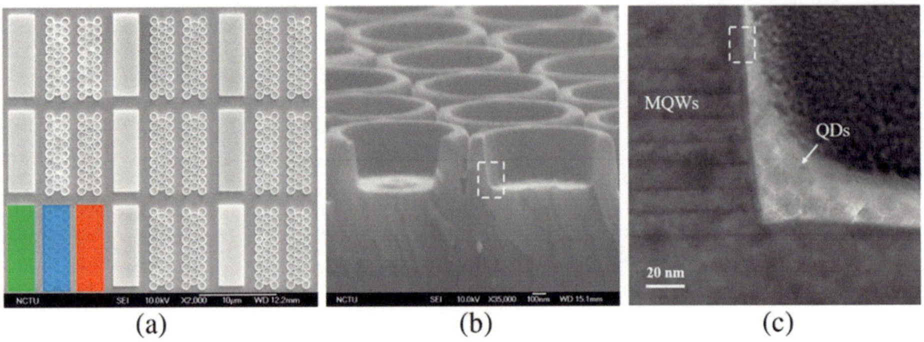

Figure 4.
(a) SEM image of RGB pixel array (top view); (b) SEM image of NR-μLED with 30° tilt angle; (c) TEM image of the contact area between MQWs and QDs [32]. Figures reproduced with permission from Optica Publishing Group.

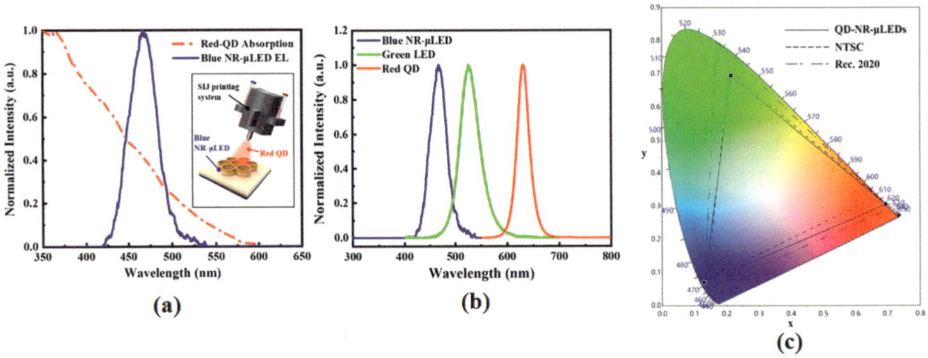

Figure 5.
(a) Absorption curve of red QDs and electroluminescence (EL) spectrum of blue NR-μLEDs (inset depicts a schematic configuration of spraying red QDs on blue NR-μLEDs using the SIJ printing system). (b) EL spectra of RGB hybrid QD-NR-μLEDs; (c) color gamut of RGB hybrid QD-NR-μLEDs, NTSC, and Rec. 2020 [32]. Figures reproduced with permission from Optica Publishing Group.

makes sure the strong coupling exists between excitons in MQWs and the absorption dipoles of QDs. Moreover, the inset of **Figure 5(a)** schematically shows the process of spraying red QDs on blue NR-μLEDs by using the SIJ printing system. **Figure 5(b)** demonstrated the EL spectra of individual RGB colors in hybrid QD-NR-μLEDs, the peak wavelengths at 467,525, and 630 nm, respectively. Due to the narrow EL spectra, the fabricated QD-NR-μLEDs can achieve a wide color gamut, which overlaps the NTSC space at approximately 104.8%, and overlaps Rec. 2020 space at about 78.2%.

In recent years, inkjet printing combined with photolithography, namely photolithography-inkjet printing (PHO-IJP) is usually used to fabricate a single pixel of conventional semiconductor quantum dots, with pixel pits on a substrate being created by photolithography followed by filling them with ink via inkjet printing [33]. In 2022, Bai et al. proposed “Interface Engineering-Inkjet Printing-Plasma Etching (IE-IJP-PE)” to fabricate large-area μPeLEDs with microscale and self-emissive pixels [34]. To achieve full-color display, $CsPbCl_{1.56}Br_{1.44}$ (blue), $CsPbBr_3$ (green), and $CsPbBrI_2$ (red) PeQD cyclohexylbenzene solutions were used as RGB inks. The micro-PeLED arrays were constructed with a structure of glass/ITO/PEDOT: PSS/PVK/SDS/RGB PeQD arrays/TPBi/LiF/Al. The fabricated substrates were ultrasonically cleaned sequentially in detergent, acetone, ethanol, and deionized water and dried. After the devices were dried, the ITO glass substrates were treated with UV ozone to remove residual organics for 15 min. The filtered PEDOT:PSS solution with a poly(tetrafluoroethylene) syringe filter (0.45 μm) was spin-coated onto an ITO glass substrate at 4000 rpm for 30 s and annealed at 140°C for 15 min. Dissolve PVK in chlorobenzene to form a 5 mg ml solution. The filtered PVK solution was spin-coated with a 0.45 μm poly(tetrafluoroethylene) syringe filter onto the PEDOT:PSS layer at 3000 rpm for 30 s and annealed at 130°C for 15 min in a glove box. SDS was then spin-coated at 3000 rpm for 30 s and heated at 100°C for 15 min. The RGB PeQD ink was printed on the hole transport layer in air by an IJP process and then treated by plasma cleaning for 15 s to destroy the excess hole transport layer and reduce the leakage current of the micro-PeLED array. Finally, the samples with RGB PeQD arrays were transferred to an interconnected high-vacuum deposition system through thermal evaporation to complete the device. The color gamut of RGB μPeLEDs covers 135% of NTSC, which is much wider than full-color μLEDs fabricated from red and green CdSe-based QDs. In addition, the red μPeLED has a maximum EQE of 0.832%, which is higher than its InGaN-based and AlInGaP-based counterparts.

As mentioned earlier, QD printing technology has become increasingly mature. However, inkjet printing is difficult to achieve large-area production. Colloidal quantum dots exhibit unique optical properties derived from quantum confinement effects that make them suitable for use as color-conversion layers for μLEDs [35]. PRs can be used to combine with quantum dots to form QDPRs after surface modification. By changing the parameters, QDPR can freely control thickness and size, while retaining the advantages of lithography. This method provides a cost-effective and practical solution for the development of large-area, high-resolution fabrication of full-color μLEDs for display applications.

In 2020, Chen et al. demonstrated a full-color μLED display with high color stability using semipolar (20–21) InGaN LEDs and quantum-dot photoresist [36]. To overcome the expensive and not desirable mass production of the conventional way, an innovative orientation-controlled epitaxy (OCE) process with semipolar GaN material selectively grown directly on the standard sapphire wafer was used in this work. The μLED array process started with depositing a transparent conducting oxide (TCO) layer, then, a p-type ohmic contact was formed by annealing at 450°C with the ambient atmospheric conditions. Next, in order to form a 1 μm depth mesa etch and etch the TCO film, inductively coupled plasma-reactive ion etching (ICP-RIE) and an HCl solution were used respectively in this research. Afterward, by using electron beam evaporation, Ti/Al/Ti/Au layers were deposited as the n-type electrode. After that, with the help of plasma-enhanced chemical vapor deposition, a 200 nm thick SiO_2 passivation layer was deposited. Eventually, followed by ICP-RIE, the μLED array was completed by the via-hole process. After the μLED array process, as shown in **Figure 6(a)**, the Ni/Au (p-electrode metal) lines were deposited on the flattened surface to link each chip, as shown in **Figure 6(b)**. Then, by using lithography process, the gray photoresist, red QDPR, green QDPR, and transparent PR were fabricated sequentially to form a color pixel on a highly-transparent glass substrate with 0.7 mm thickness, as shown in **Figure 6(c)**. Finally, in **Figure 6(d)**, an aligner and UV resin are used to stick the color pixel array on the glass. The gray PR mold can attain a higher height and provide higher reflectivity, which can reduce the cross-talk effect among pixels and enhance output intensity by inside reflection compared to the

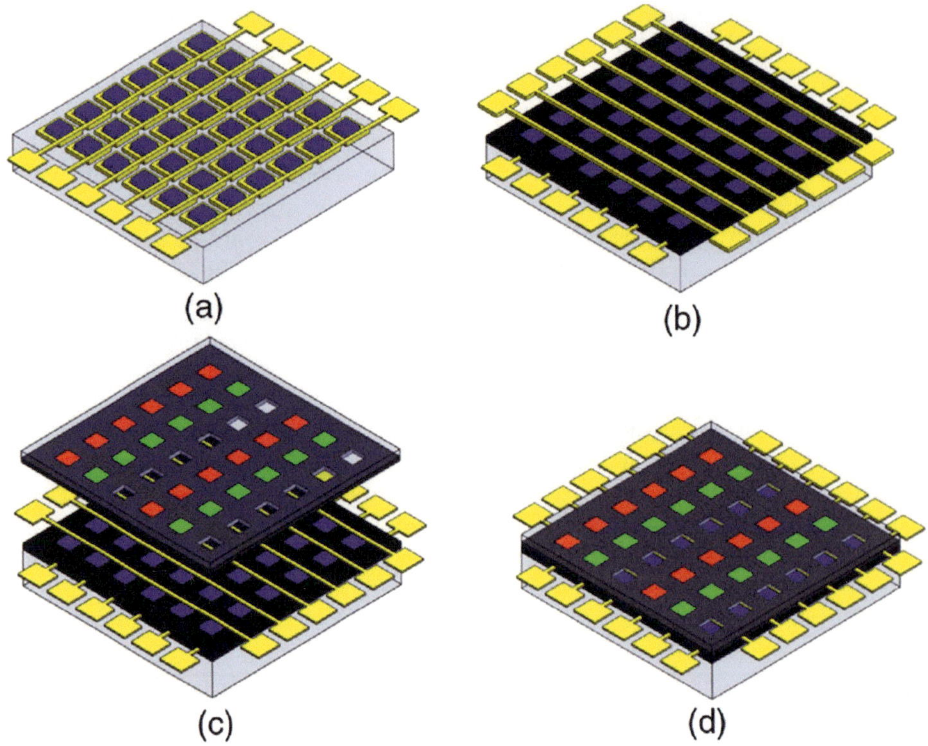

Figure 6.
Process flow for the fabrication of a full-color RGB pixel array. (a) μLED array process. (b) Black PR matrices and p-electrode metal lines. (c) Red, green, and blue (transparent) pixel lithography process. (d) Color pixel bonding [36]. Figures reproduced with permission from Optica Publishing Group.

black PR. Besides, the semipolar μLED shows a stabilizing wavelength shift of 3.2 nm, while the c-plane μLED's shift is 13.0 nm. Above all, the RGB pixels present a wide color gamut of 114.4% NTSC and 85.4% Rce.2020, showing great promise for display applications. The research also demonstrated a color-conversion layer consisting of QDPR is capable of the common lithography process, which is suitable for large-scale manufacturing.

Recently, μLED all-rounder displays based on silicon backplanes have also attracted the attention of researchers. In 2020, Kawanishi et al. reported a silicon-based full-color micro-LED display, called "Silicon Display" [37]. To fabricate the display array, a p-electrode layer is first formed on the p-GaN layer of the LED epitaxial wafer. The epitaxial layer was then etched down to the n-GaN layer by photolithographic patterning using ICP to form the mesa structure, and the n-GaN layer was exposed to form the n-electrode. Each pixel is defined by a groove down to the sapphire substrate formed by another ICP etch. **Figure 7(a)** shows a schematic diagram of a single pixel. Red and green sub-pixels are formed by exciting the QD color converters with blue LEDs. The quantum dot material is deposited on the surface of the device by a photolithography process after being mixed with a photoresist. A full-color silicon display with a resolution of 1053 ppi and 352 × 198 pixels, each of which is 24 μm in size, is realized. **Figure 7(b)** shows a photo of the overall display. As can be seen from the photo, each QD layer was successfully formed on each sub-pixel through the photolithography process.

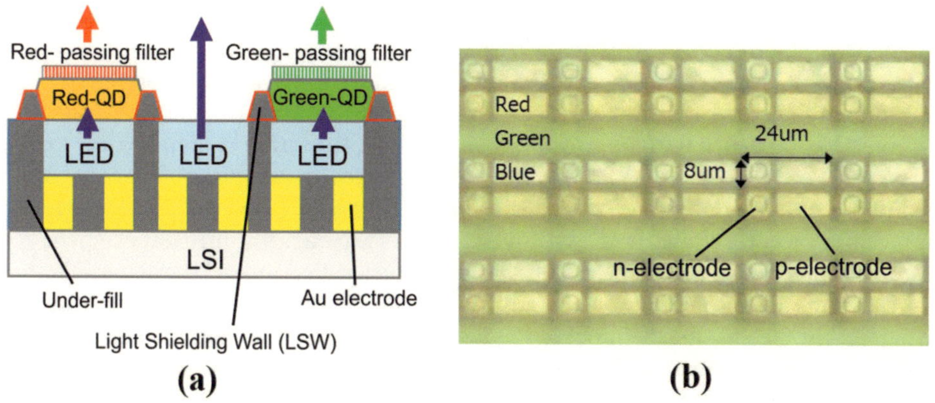

Figure 7.
(a) Schematic cross-section of a single pixel of silicon display. (b) Optical micrograph of 1053 ppi micro-LED array during fabrication after forming red and green QDs [37]. Figures reproduced with permission from the SOCIETY FOR INFORMATION DISPLAY.

2.4 VLC applications with QD-based color-conversion μLEDs

Incorporating a color conversion layer can allow μLEDs to simultaneously achieve full-color display and high-speed modulation. However, the modulation bandwidth of this class of devices is significantly limited by the long response times of color-converting materials [38]. The traditional phosphor material used as the color conversion layer is yttrium aluminum garnet (YAG) phosphor Y_3–xAl_5O_{12}:xCe^{3+} (YAG:Ce) [39], which has a critical limitation for VLC applications due to the slow phosphor conversion process caused by the long excited-state lifetimes [40], on the order of microseconds. The modulation bandwidth of phosphors is typically only a few MHz [41]. To overcome the bottleneck in response speed, organic materials, such as BODIPY, MEH-PPV and BBEHP-PPV have been recently used as potential candidates for color converters for VLCs due to their visible light emission, high PLQY, direct radiative recombination, and ease of integration with nitride-based semiconductors [42]. However, their excited state lifetimes are still very long. Therefore, developing light-converter phosphor materials with fast decay and high efficiency (that is, short radiative lifetime and high brightness) remains a major challenge for VLC and solid-state lighting (SSL) applications.

As previously introduced, quantum dot (QD)-based color converters have very promising applications, however, the modulation bandwidth of conventional CdSe/ZnS QDs is limited to ~3 MHz, which is much lower than the requirement of VLC [43]. Lead halide perovskite QDs exhibit high PLQY (≥70%) and relatively short PL lifetimes [27]. In 2018, Shi et al. reported an all-inorganic white light system for VLC [44]. The system uses blue GaN-based μLEDs as the excitation light source and inorganic yellow-emitting $CsPbBr_{1.8}I_{1.2}$ perovskite quantum dots (YQDs) as the color conversion layer. The maximum modulation bandwidth of the packaged 80 μm × 80 μm blue-emitting μLED is about 160 MHz, and the peak emission wavelength is about 445 nm. Maximum −3 dB E-O modulation bandwidths of ~73 and ~85 MHz were achieved for perovskite quantum dots and white light systems combining μLED and perovskite quantum dots, respectively. In addition, based on the high bandwidth white light system, the real-time data rate is 300 Mbps using no return Zero-On–Off Keying (NRZ-OOK) modulation scheme.

Most GaN-based μLEDs are typically grown on (0001) "polar" c-plane sapphire substrates, which leads to strong Stark effect QCSE effect (QCSE), which, in addition to leading to a drop in efficiency, will limit the modulation bandwidth of μLEDs [45]. The spontaneous polarization of GaN is responsible for the QCSE due to the highest symmetry compatible with its structure. Meanwhile, the strain caused by the lattice mismatch between InxGa1 − xN and GaN also produces polarization. These internal polarization fields along the c-plane will lead to band tilting, separating the wave functions of electrons and holes. QCSE also causes wavelength shift and efficiency drop with increasing injection current density. The applications of the semipolar (20–21) and (20–2–1) epitaxial structures have been proven to effectively suppress the effects of QCSE [46]. Semipolar devices enable higher modulation bandwidths due to weak polarization fields and flat energy gap distributions, which lead to larger electron–hole wavefunction overlap reducing carrier lifetimes [47]. Further, Zhao et al. revealed that faster carrier transport in semipolar devices also contributes to the weaker phase-space filling effect, which was determined for the low-droop phenomenon in semipolar LEDs because of small QCSE and short carrier lifetimes using the consistency between theoretical and experimental results [48]. The aforementioned advantages imply that a semipolar LED is capable of simultaneously achieving high modulation speed and maintaining high efficiency with increasing injected current owing to low droop performance.

In 2020, Huang Chen et al. realized long-wavelength (initial wavelength 540 nm) InGaN/GaN VLC-LEDs with high 3 dB bandwidth using semipolar epitaxy and μLED structures [49]. The epitaxial process of semipolar (20–21) GaN on a (22–43) PSS is carried out through a low-pressure metalorganic chemical vapor deposition (MOCVD). In addition, the passivation layer of aluminum oxide (Al_2O_3) was grown to repair sidewall defects. Some previous studies have stated that the influence of sidewall defects increases as the chip size decreases [50, 51]. In particular, when the LED device achieves a micrometer scale, traditional passivation methods, such as the PECVD process, are no longer useful owing to the large leakage current of the μLED device. ALD dielectric thin films have been regarded as an effective passivation technique in the μLED area [52]. The semipolar device had a shorter lifetime because of the weak polarization-related electric field and large overlap of the electron–hole wave function, which yielded a faster carrier recombination lifetime [53]. The QCSE reduction in the semipolar device yielded faster carrier transport and a shorter recombination lifetime, resulting in a weaker phase-space filling effect. Therefore, the semipolar μLED can achieve a high modulation bandwidth owing to its faster carrier recombination lifetime.

The outstanding performance of semipolar μLEDs in display and communication has also led researchers to combine it with QDs to make high-performance full-color display devices with VLC potential. In 2021, Singh et al. proposed a flexible white-light system for high-speed VLC applications [54]. The white-light system fabrication process is shown in **Figure 8**. The system consists of nanostructured green $CsPbBr_3$ PQD paper, red CdSe QD paper, and semipolar blue micro-LEDs. Regarding the production of green $CsPbBr_3$ PQD paper, firstly, a solution of $CsPbBr_3$ quantum dots were prepared using the hot injection method, and then the solution was added to the cellulose nanocrystal (CNC) suspension, and the mixed solution was filtered through a filter membrane using a vacuum pump device. The PQD paper in nanostructured form is then separated from the filter membrane. The QD paper produced by this method has strong mechanical strength and flexibility to be used with flexible systems. In addition, the PQD paper fabricated by this method has nanostructures, which should provide a strong quantum confinement effect to increase

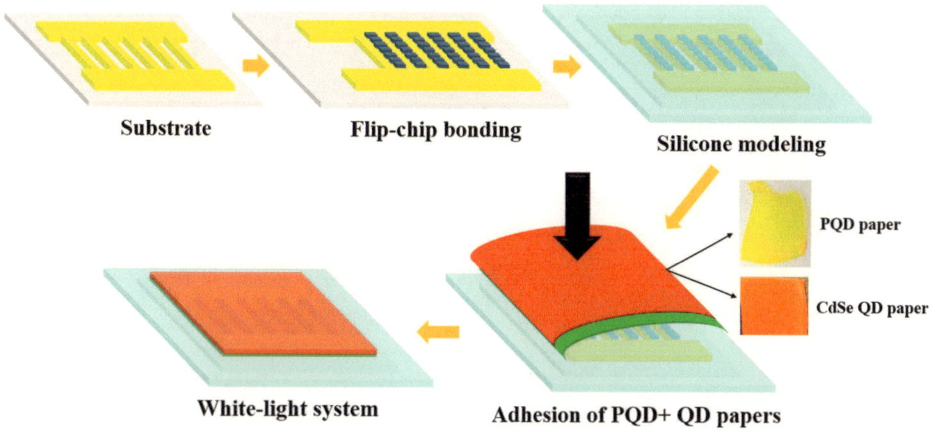

Figure 8.
Fabrication of white-light system [54]. Figures reproduced with permission from Optica Publishing Group.

the probability of carrier recombination. The flexible μLED was fabricated using a polyimide (PI) substrate covered with copper-foil shielding tape, where the latter was subjected to photolithography and wet etching to establish electrical conduction. The μLED flip-chip was bonded on the PI substrate using a silicone-based electrically conductive anisotropic adhesive to maintain the electrical conductivity between the chip metal contact and the AuSn solder on the substrate; this increased the system flexibility. For the color converter, $CsPbBr_3$ PQD paper and CdSe QD paper were prepared. These papers were adhered to the top of the μLED with flexible substrate using an adhesive to achieve white-light system.

Semipolar μLEDs had a narrow FWHM and was therefore responsible for the delivery of pure emitted light, matching colors, and a wide color gamut. **Figure 9(a)** shows the color performance of the white-light system created using the semipolar blue μLED with PQD paper and CdSe QD paper under driving conditions from 10 to 1200 A/cm^2. The white-light system demonstrated a wide color gamut, achieving 98.7% of the NTSC and 91.1% Rec. 2020 of the CIE 1931. The color gamut of the white-light system remained almost unchanged with increasing injection current density owing to the wavelength stability.

The average PL lifetimes calculated for the PQD paper and CdSe QD paper were 5.92, and 12.88 ns, as shown in **Figure 9(b)**. The PQD paper had a shorter carrier lifetime than those reported in other studies, while also being considerably shorter than those of phosphors microsecond to millisecond range. This shorter carrier lifetime is attributable to the quantum confinement effect, which yields faster radiative recombination. However, the CdSe QD-paper carrier recombination lifetime is insufficient to independently achieve high bandwidth for VLC applications. Therefore, semipolar μLEDs and PQDs have considerable potential for VLC applications. The PQD-film bandwidth was found to be 111 MHz, as shown in **Figure 9(c)**; the PQD-based white-light system also displayed a frequency bandwidth of 95.5 MHz at a 113 mA injection current. Hence, the high bandwidth of the PQD paper is suitable for achieving high-speed VLC. This outcome implies that the nanostructure has a higher recombination rate than the bulk and hence a higher modulation bandwidth.

Although PQDs have significant advantages over traditional color conversion materials, they also have some drawbacks. They have, for example, exhibited vulnerability under ambient conditions, particularly in the case of red-emitting PQDs that

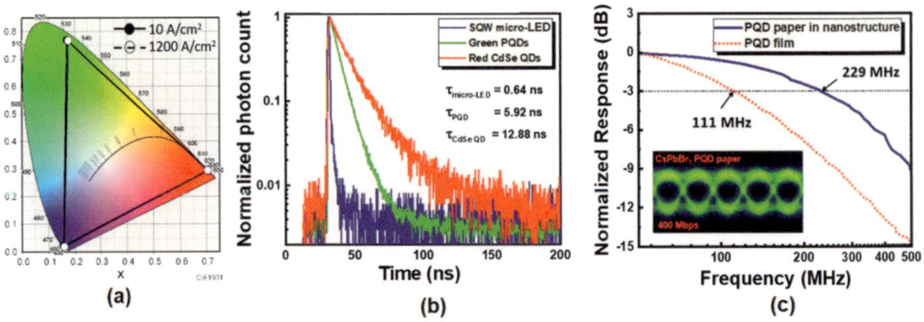

Figure 9.
(a) Color gamut of white-light system according to CIE 1931 color space under various current densities. (b) TRPL curves for semipolar μLED and PQD and CdSe QD papers. (c) Comparison of bandwidth of PQD paper in nanostructure with that of PQD film, Inset: eye diagram for PQD paper [54]. Figures reproduced with permission from Optica Publishing Group.

contain iodine [55]. The application bottleneck of PQDs is long-term stability. It degrades rapidly when exposed to the environment. Water vapor, oxygen, high temperature, and light irradiation cause alteration to the crystal structures of PQDs, typically resulting in photoluminescence (PL) quenching [56]. These four factors coexist in the environment, so their effects are difficult to distinguish from each other [57]. In terms of immediate optical performance, most of the crystal structure changes are negative, but a few of them are positive. Therefore, the high-stability PQDs color conversion layer made by the new encapsulation method will have good application prospects in the fields of display and communication. Embedding QDs in organic polymers can significantly extend their lifetimes, as the polymers ensure a hermetic seal from air [58, 59]. In the current study, the organic shell is still not perfect because the polymer cannot withstand UV or blue illumination from the excitation light source for long periods of time [60]. Thus, shells composed of inorganic substances have been favored since recently, including Al_2O_3, ZrO_2, and anodized aluminum templates [61, 62]. Although these inorganic shells can withstand blue light and UV irradiation better than organic shells, the porous structures are not as water/O_2-proof as the organic ones because the pore structures remain. In 2021, Lin et al. reported an inorganic encapsulation of mesoporous SiO_2, in conjunction with a high-temperature sintering synthesis process under an inert atmosphere [58]. This synthesis process is compatible with various halide contents, yielding PeNCs sealed in SiO_2 particles that emit PL emission covering the entire visible range from 420 to 700 nm. The PeNCs–SiO_2 sample showed remarkable stability after undergoing aging tests under various exaggerated stresses as well as mitigated thermal quenching during thermal cycling. The PeNCs–SiO_2 can be blended into a photoresist and remains luminous during the development procedure, which is compatible with the photolithography process; this facilitates the mass production of color conversion layers. This robust encapsulation also has great application value for full-color display or visible light communication based on quantum dot-based μLED.

In 2021, Wu et al. reported a PNC–μLED device for a full-color display that is developed using a semipolar (20–21) blue μLED array with green-emitting $CsPbBr_3$ and red-emitting $CsPbBrI_2$ PNCs [63]. They encapsulated the PQDs in an all-inorganic SiO_2 shell, which significantly improved the stability of the color conversion layer. Regarding the fabrication of $CsPbBrI_2$-SiO_2, the precursor solution was first prepared by mixing the precursor salt with MCM-41 molecular sieves and dispersing the mixture in 25 ml of purified water. Next, the precursor solution was sonicated for 20 min and vigorously stirred for 10 min to improve dispersion. The precursor solution was transferred into a crucible and placed in a tube furnace filled with high-purity Ar gas. The temperature of the tube furnace was raised to 200°C and kept for 1 h to evaporate the water. Then, sintering was performed continuously for 30 min at a temperature of 750°C with an Ar flow rate of 15 ml/min. Subsequently, the samples were cooled to room temperature under Ar protection, during which their color gradually changed and finally crystallized into PNCs. During cooling, the perovskite forms inside the SiO_2, which constrains the PNCs inside that facilitate the formation of crystal phase, thereby ensuring the high stability of the red-emitting PNCs. **Figure 10(a)** illustrates

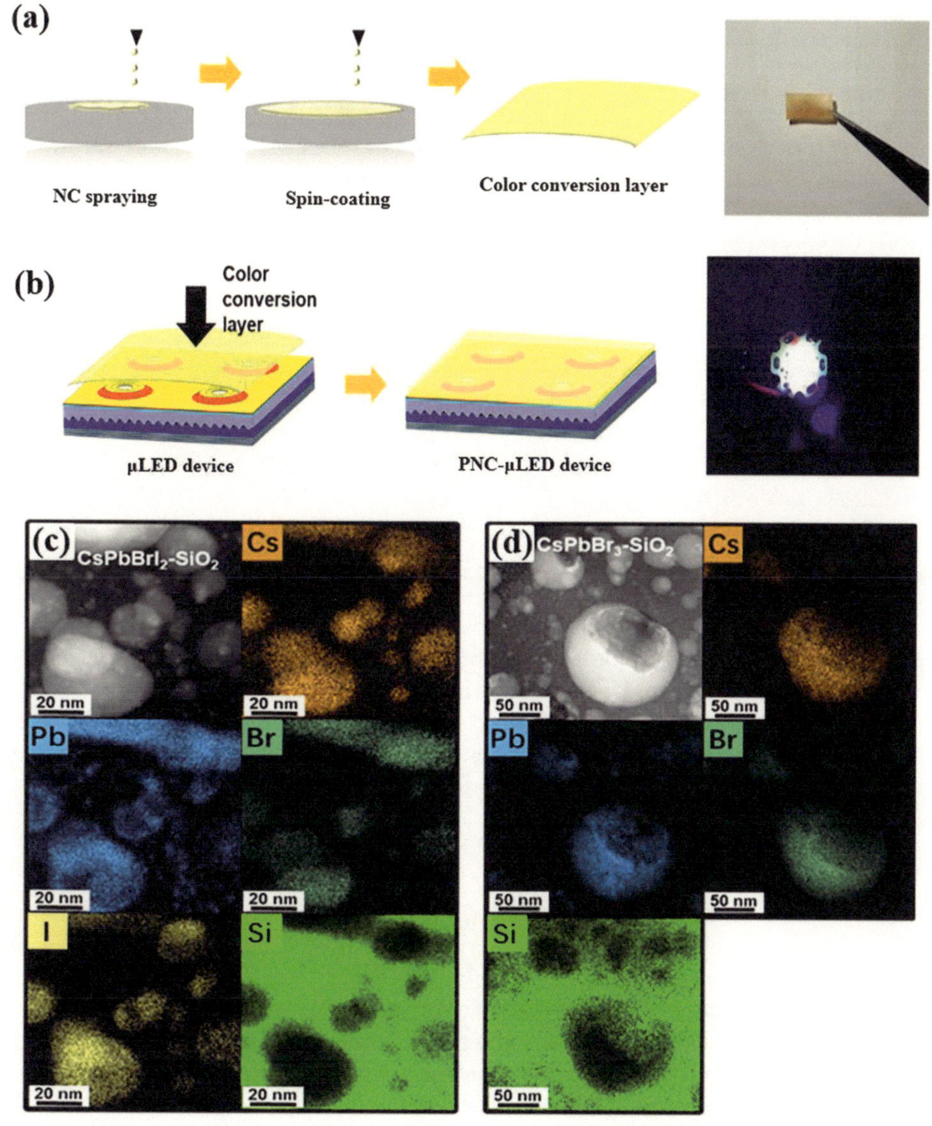

Figure 10.
(a) The fabrication process and a photo of the color conversion layer; and (b) the proposed PNC-µLED device. EDS element maps of the (a) $CsPbBrI_2$-SiO_2 and (b) $CsPbBr_3$-SiO_2 PNCs [63]. Figures reproduced with permission from Optica Publishing Group.

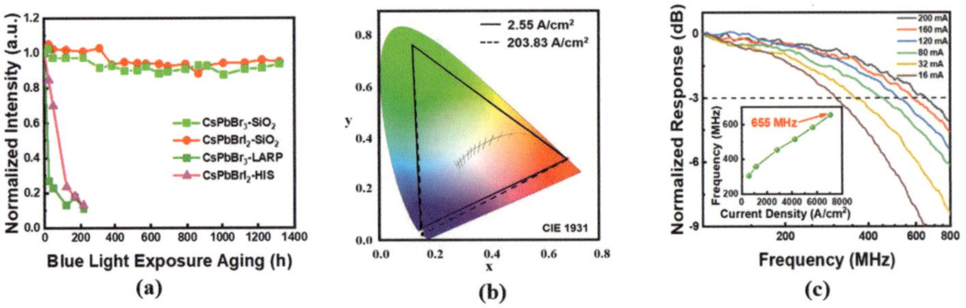

Figure 11.
(a) The normalized intensity of the PL properties of the red and green PNCs, as well as the solution-processed samples under blue light exposure. (b) Color gamut of the PNC-μLED under different current densities. (c) Frequency response for the PNC-μLED [63]. Figures reproduced with permission from Optica Publishing Group.

the fabrication detail and presents a photo of the as-fabricated films under natural light, which are composed of $CsPbBr_3$-SiO_2 and $CsPbBrI_2$-SiO_2. During this process, $CsPbBrI_2$-SiO_2 and $CsPbBr_3$-SiO_2 were mixed with toluene to make a red and green mixture. The EVA polymer was dissolved into the mixture and heated to 50°C with stirring. The mixture was then spin-coated (1800 rad/s) on a glass substrate to obtain a thin film of uniform thickness. In **Figure 10(b)**, the color conversion layer is combined with the semipolar μLED, achieving the white-light PNC-μLED device for the backlight. The fine structure of both PNC samples is also revealed from the EDS element maps, as illustrated in **Figure 10(c)** and **(d)**. These maps indicate that the spatial distribution of the elements within the PNCs is highly similar; they are confined to approximately circular regions surrounded by areas of SiO_2. This further confirms the encapsulation of the PNCs inside SiO_2 shells.

Figure 11(a) shows the normalized intensity of the PL properties of the red and green PNCs, as well as the solution-processed samples under blue light exposure. The SiO_2-embedded samples exhibit remarkable stabilities, showing no degradation under blue irradiation. **Figure 11(b)** shows the frequency response of the PNC-μLED, the highest 3-dB bandwidth was measured as 655 MHz, corresponding to an injection current of 200 mA. **Figure 11(c)** presents the performance of the PNC-μLED for the display backlight application under different current densities between 2.55 and 203.83 A/cm^2. Because of its narrow EL spectrum, the RGB pixel assembled from the semipolar μLEDs and PNCs exhibited a wide color gamut of 127.23% of the NTSC and 95.00% of the Rec. 2020.

3. Conclusions

QDs-based μLED provide superior display performance and are a promising platform for VLC systems. In this article, we comprehensively review recent progress in QD-based μLED devices. It includes the research status of various QDs and white LEDs based on QDs color conversion layers. The fabrication of QD-based high-resolution full-color μLEDs is also discussed. Including charge-assisted layer-by-layer (LbL), aerosol jet printing, and super inkjet printing methods to fabricate QD-based μLEDs. The use of quantum dot photoresist in combination with semipolar μLEDs is also described. Finally, we discuss the research of QD-based μLEDs for visible light communication, which allows a single device to be used for both display and high-speed communication, enhancing the versatility of μLEDs. Advances in the development of QD-based μLEDs are expected to make this display technology ubiquitous in the near future. Recent breakthroughs in QDs and LEDs will provide a promising outlook for future demand in the semiconductor industry.

Acknowledgements

Additionally, we thank Xiangshu Lin for their contributions to the investigation.

Funding

This research was supported by the National Natural Science Foundation of China (62274138, 11904302), Science and Technology Plan Project in Fujian Province of China (2021H0011), Major Science and Technology Project of Xiamen, China (3502Z20191015).

Author details

Tingzhu Wu[1,2], Tingwei Lu[1], Yen-Wei Yeh[3], Zhong Chen[1,2] and Hao-Chung Kuo[3,4*]

1 School of Electronic Science and Engineering, Fujian Engineering Research Center for Solid-State Lighting, Xiamen University, Xiamen, China

2 Innovation Laboratory for Sciences and Technologies of Energy Materials of Fujian Province (IKKEM), Xiamen, China

3 Department of Photonics, Graduate Institute of Electro-Optical Engineering, College of Electrical and Computer Engineering, National Yang Ming Chiao Tung University, Hsinchu, Taiwan

4 Semiconductor Research Center, Hon Hai Research Institute, Taipei, Taiwan

*Address all correspondence to: hckuo@faculty.nctu.edu.tw

References

[1] Alivisatos P. The use of nanocrystals in biological detection. Nature Biotechnology. 2004;**22**(1):47-52. DOI: 10.1038/nbt927

[2] Tian PF, McKendry JJD, Gu ED, Chen ZZ, Sun YJ, Zhang GY, et al. Fabrication, characterization and applications of flexible vertical InGaN micro-light emitting diode arrays. Optics Express. 2016;**24**(1):699-707. DOI: 10.1364/oe.24.000699

[3] Liu ZJ, Zhang K, Liu YB, Yan SW, Kwok HS, Deen J, et al., editors. Fully multi-functional GaN-based micro-LEDs for 2500 PPI micro-displays, temperature sensing, light energy harvesting, and light detection. In: 64th IEEE Annual International Electron Devices Meeting (IEDM); December 01-05, 2018. New York, San Francisco, CA: IEEE; 2018

[4] Singh KJ, Huang YM, Ahmed T, Liu AC, Chen SWH, Liou FJ, et al. Micro-LED as a promising candidate for high-speed visible light communication. Applied Sciences—Basel. 2020;**10**(20):7384. DOI: 10.3390/app10207384

[5] Khan LU. Visible light communication: Applications, architecture, standardization and research challenges. Digital Communications and Networks. 2017;**3**(2):78-88. DOI: 10.1016/j.dcan.2016.07.004

[6] Chow CW, Yeh CH, Liu YF, Liu Y. Improved modulation speed of LED visible light communication system integrated to main electricity network. Electronics Letters. 2011;**47**(15):867-U1954. DOI: 10.1049/el.2011.0422

[7] Cho J, Park JH, Kim JK, Schubert EF. White light-emitting diodes: History, progress, and future. Laser & Photonics Reviews. 2017;**11**(2):1600147. DOI: 10.1002/lpor.201600147

[8] Bulashevich KA, Karpov SY. Impact of surface recombination on efficiency of III-nitride light-emitting diodes. Physica Status Solidi-Rapid Research Letters. 2016;**10**(6):480-484. DOI: 10.1002/pssr.201600059

[9] Hwang D, Mughal A, Pynn CD, Nakamura S, DenBaars SP. Sustained high external quantum efficiency in ultrasmall blue III-nitride micro-LEDs. Applied Physics Express. 2017;**10**(3):032101. DOI: 10.7567/apex.10.032101

[10] Wang XC, Bao Z, Chang YC, Liu RS. Perovskite quantum dots for application in high color gamut backlighting display of light-emitting diodes. ACS Energy Letters. 2020;**5**(11):3374-3396. DOI: 10.1021/acsenergylett.0c01860

[11] Lin CH, Verma A, Kang CY, Pai YM, Chen TY, Yang JJ, et al. Hybrid-type white LEDs based on inorganic halide perovskite QDs: Candidates for wide color gamut display backlights. Photonics Research. 2019;**7**(5):579-585. DOI: 10.1364/prj.7.000579

[12] Lin CC, Liu RS. Advances in phosphors for light-emitting diodes. Journal of Physical Chemistry Letters. 2011;**2**(11):1268-1277. DOI: 10.1021/jz2002452

[13] Huang YM, Singh KJ, Hsieh TH, Langpoklakpam C, Lee TY, Lin CC, et al. Gateway towards recent developments in quantum dot-based light-emitting diodes. Nanoscale. 2022;**14**(11):4042-4064. DOI: 10.1039/d1nr05288h

[14] Sapsford KE, Pons T, Medintz IL, Mattoussi H. Biosensing

with luminescent semiconductor quantum dots. Sensors. 2006;**6**(8):925-953. DOI: 10.3390/s6080925

[15] Araki Y, Ohkuno K, Furukawa T, Saraie J. Green light emitting diodes with CdSe quantum dots. Journal of Crystal Growth. 2007;**301**:809-811. DOI: 10.1016/j.jcrysgro.2006.11.105

[16] Du JH, Wang CL, Xu XJ, Wang ZY, Xu SH, Cui YP. Assembly of light-emitting diode based on hydrophilic CdTe quantum dots incorporating dehydrated silica gel. Luminescence. 2016;**31**(2):419-422. DOI: 10.1002/bio.2976

[17] Morris-Cohen AJ, Donakowski MD, Knowles KE, Weiss EA. The effect of a common purification procedure on the chemical composition of the surfaces of CdSe quantum dots synthesized with Trioctylphosphine oxide. Journal of Physical Chemistry C. 2010;**114**(2):897-906. DOI: 10.1021/jp909492w

[18] Stan CS, Secula MS, Sibiescu D. Highly luminescent polystyrene embedded CdSe quantum dots obtained through a modified colloidal synthesis route. Electronic Materials Letters. 2012;**8**(3):275-281. DOI: 10.1007/s13391-012-1108-0

[19] Soheyli E, Ghaemi B, Sahraei R, Sabzevari Z, Kharrazi S, Amani A. Colloidal synthesis of tunably luminescent AgInS-based/ZnS core/shell quantum dots as biocompatible nano-probe for high-contrast fluorescence bioimaging. Materials Science & Engineering C-Materials for Biological Applications. 2020;**111**:110807. DOI: 10.1016/j.msec.2020.110807

[20] Reifsnyder DC, Ye XC, Gordon TR, Song CY, Murray CB. Three-dimensional self-assembly of chalcopyrite copper indium Diselenide nanocrystals into oriented films. ACS Nano. 2013;**7**(5):4307-4315. DOI: 10.1021/nn4008059

[21] Chen B, Li DY, Wang F. InP quantum dots: Synthesis and lighting applications. Small. 2020;**16**(32):2002454. DOI: 10.1002/smll.202002454

[22] Brus LE. Electron electron and electron-hole interactions in small semiconductor crystallites—The size dependence of the lowest excited electronic state. Journal of Chemical Physics. 1984;**80**(9):4403-4409. DOI: 10.1063/1.447218

[23] Chen O, Wei H, Maurice A, Bawendi M, Reiss P. Pure colors from core-shell quantum dots. MRS Bulletin. 2013;**38**(9):696-702. DOI: 10.1557/mrs.2013.179

[24] Zhang ZL, Liu D, Li DZ, Huang KK, Zhang Y, Shi Z, et al. Dual emissive Cu:InP/ZnS/InP/ZnS nanocrystals: Single-source "greener" emitters with flexibly Tunable emission from visible to near-infrared and their application in white light-emitting diodes. Chemistry of Materials. 2015;**27**(4):1405-1411. DOI: 10.1021/cm5047269

[25] Milstein TJ, Kroupa DM, Gamelin DR. Picosecond quantum cutting generates photoluminescence quantum yields over 100% in ytterbium-doped CsPbCl3 nanocrystals. Nano Letters. 2018;**18**(6):3792-3799. DOI: 10.1021/acs.nanolett.8b01066

[26] Schmidt LC, Pertegas A, Gonzalez-Carrero S, Malinkiewicz O, Agouram S, Espallargas GM, et al. Nontemplate synthesis of CH3NH3PbBr3 perovskite nanoparticles. Journal of the American Chemical Society. 2014;**136**(3):850-853. DOI: 10.1021/ja4109209

[27] Protesescu L, Yakunin S, Bodnarchuk MI, Krieg F, Caputo R, Hendon CH, et al. Nanocrystals of cesium Lead halide perovskites (CsPbX3, X = Cl, Br, and I): Novel optoelectronic materials showing bright emission with wide color gamut. Nano Letters. 2015;**15**(6):3692-3696. DOI: 10.1021/nl5048779

[28] Wang HC, Bao Z, Tsai HY, Tang AC, Liu RS. Perovskite quantum dots and their application in light-emitting diodes. Small. 2018;**14**(1):1702433. DOI: 10.1002/smll.201702433

[29] Kong CY, Lin CH, Lin CH, Li TY, Chen SWH, Tsai CL, et al. Highly efficient and stable white light-emitting diodes using perovskite quantum dot paper. Advanced. Science. 2019;**6**(24):1902230. DOI: 10.1002/advs.201902230

[30] Han HV, Lin HY, Lin CC, Chong WC, Li JR, Chen KJ, et al. Resonant-enhanced full-color emission of quantum-dot-based micro LED display technology. Optics Express. 2015;**23**(25):32504-32515. DOI: 10.1364/oe.23.032504

[31] Lin HY, Sher CW, Hsieh DH, Chen XY, Chen HMP, Chen TM, et al. Optical cross-talk reduction in a quantum-dot-based full-color micro-light-emitting-diode display by a lithographic-fabricated photoresist mold. Photonics Research. 2017;**5**(5):411-416. DOI: 10.1364/prj.5.000411

[32] Chen SWH, Shen CC, Wu TZ, Liao ZY, Chen LF, Zhou JR, et al. Full-color monolithic hybrid quantum dot nanoring micro light-emitting diodes with improved efficiency using atomic layer deposition and nonradiative resonant energy transfer. Photonics Research. 2019;**7**(4):416-422. DOI: 10.1364/prj.7.000416

[33] Xuan TT, Shi SC, Wang L, Kuo HC, Xie RJ. Inkjet-printed quantum dot color conversion films for high-resolution and full-color micro light-emitting diode displays. Journal of Physical Chemistry Letters. 2020;**11**(13):5184-5191. DOI: 10.1021/acs.jpclett.0c01451

[34] Bai WH, Xuan TT, Zhao HY, Shi SC, Zhang XY, Zhou TL, et al. Microscale perovskite quantum dot light-emitting diodes (micro-PeLEDs) for full-color displays. Advanced Optical Materials. 2022;**10**(12):2200087 DOI: 10.1002/adom.202200087

[35] Erdem T, Demir HV. Color science of nanocrystal quantum dots for lighting and displays. Nano. 2013;**2**(1):57-81. DOI: 10.1515/nanoph-2012-0031

[36] Chen SWH, Huang YM, Singh KJ, Hsu YC, Liou FY, Song J, et al. Full-color micro-LED display with high color stability using semipolar (20-21) InGaN LEDs and quantum-dot photoresist. Photonics Research. 2020;**8**(5):630-636. DOI: 10.1364/prj.388958

[37] Kawanishi H, Onuma H, Maegawa M, Kurisu T, Ono T, Akase S, et al. High-resolution and high-brightness full-colour "silicon display" for augmented and mixed reality. Journal of the Society for Information Display. 2021;**29**(1):57-67. DOI: 10.1002/jsid.968

[38] Xu Y, Chen J, Zhang H, Wei H, Zhou L, Wang Z, et al. White-light-emitting flexible display devices based on double network hydrogels crosslinked by YAG:Ce phosphors. Journal of Materials Chemistry C. 2020;**8**(1):247-252. DOI: 10.1039/C9TC05311E

[39] Li PP, Lu Y, Duan YM, Xu SQ, Zhang JJ. Potential application of perovskite glass material in photocatalysis field. Journal of Physical Chemistry C. 2021;**125**(4):2382-2392. DOI: 10.1021/acs.jpcc.0c11241

[40] Wang ZM, Wei ZX, Cai YT, Wang L, Li MT, Liu P, et al. Encapsulation-enabled

perovskite-PMMA films combining a Micro-LED for high-speed white-light communication. ACS Applied Materials & Interfaces. 2021;**13**(45):54143-54151. DOI: 10.1021/acsami.1c15873

[41] Gao H, Xie YY, Geng C, Xu S, Bi WG. Efficiency enhancement of quantum-dot-converted LEDs by 0D-2D hybrid scatterers. ACS Photonics. 2020;7(12):5430-5439. DOI: 10.1021/acsphotonics.0c01240

[42] Su CY, Wu YC, Cheng CH, Wang WC, Wang HY, Chen LY, et al. Color-converting violet laser diode with an ultrafast BEHP-PPV. ACS Applied Electronic Materials. 2020;**2**(9):3017-3027. DOI: 10.1021/acsaelm.0c00619

[43] Xiao X, Tang H, Zhang T, Chen W, Chen W, Wu D, et al. Improving the modulation bandwidth of LED by CdSe/ZnS quantum dots for visible light communication. Optics Express. 2016;**24**(19):21577-21586. DOI: 10.1364/OE.24.021577

[44] Mei SL, Liu XY, Zhang WL, Liu R, Zheng LR, Guo RQ, et al. High-bandwidth white-light system combining a micro-LED with perovskite quantum dots for visible light communication. ACS Applied Materials & Interfaces. 2018;**10**(6):5641-5648. DOI: 10.1021/acsami.7b17810

[45] Piprek J. Efficiency droop in nitride-based light-emitting diodes. Physica Status Solidi A—Applications and Materials Science. 2010;**207**(10):2217-2225. DOI: 10.1002/pssa.201026149

[46] Liu SG, Han SC, Xu CC, Xu HW, Wang XY, Wang D, et al. Enhanced photoelectric performance of GaN-based Micro-LEDs by ion implantation. Optical Materials. 2021;**121**:111579. DOI: 10.1016/j.optmat.2021.111579

[47] Zhao YJ, Fu HQ, Wang GT, Nakamura S. Toward ultimate efficiency: Progress and prospects on planar and 3D nanostructured nonpolar and semipolar InGaN light-emitting diodes. Advances in Optics and Photonics. 2018;**10**(1):246-308. DOI: 10.1364/aop.10.000246

[48] Fu HQ, Lu ZJ, Zhao XH, Zhang YH, DenBaars SP, Nakamura S, et al. Study of low-efficiency droop in semipolar (20(2)over-bar(1)over-bar) InGaN light-emitting diodes by time-resolved photoluminescence. Journal of Display Technology. 2016;**12**(7):736-741. DOI: 10.1109/jdt.2016.2521618

[49] Chen S-WH, Huang Y-M, Chang Y-H, Lin Y, Liou F-J, Hsu Y-C, et al. High-bandwidth green semipolar (20-21) InGaN/GaN micro light-emitting diodes for visible light communication. ACS Photonics. 2020;7(8):2228-2235. DOI: 10.1021/acsphotonics.0c00764

[50] Kou JQ, Shen CC, Shao H, Che JM, Hou X, Chu CS, et al. Impact of the surface recombination on InGaN/GaN-based blue micro-light emitting diodes. Optics Express. 2019;**27**(12):A643-AA53. DOI: 10.1364/oe.27.00a643

[51] Yang W, Zhang SL, McKendry JJD, Herrnsdorf J, Tian PF, Gong Z, et al. Size-dependent capacitance study on InGaN-based micro-light-emitting diodes. Journal of Applied Physics. 2014;**116**(4):044512. DOI: 10.1063/1.4891233

[52] Wong MS, Hwang D, Alhassan AI, Lee C, Ley R, Nakamura S, et al. High efficiency of III-nitride micro-light-emitting diodes by sidewall passivation using atomic layer deposition. Optics Express. 2018;**26**(16):21324-21331. DOI: 10.1364/oe.26.021324

[53] Monavarian M, Rashidi A, Aragon A, Oh SH, Nami M, Denbaars SP, et al.

Explanation of low efficiency droop in semipolar (20(21)over-bar) InGaN/GaN LEDs through evaluation of carrier recombination coefficients. Optics Express. 2017;**25**(16):19343-19353. DOI: 10.1364/oe.25.019343

[54] Singh KJ, Fan XT, Sadhu AS, Lin CH, Liou FJ, Wu TZ, et al. CsPbBr3 perovskite quantum-dot paper exhibiting a highest 3 dB bandwidth and realizing a flexible white-light system for visible-light communication. Photonics Research. 2021;**9**(12):2341-2350. DOI: 10.1364/prj.434270

[55] Wei Y, Cheng Z, Lin J. An overview on enhancing the stability of lead halide perovskite quantum dots and their applications in phosphor-converted LEDs. Chemical Society Reviews. 2019;**48**(1):310-350. DOI: 10.1039/C8CS00740C

[56] Shangguan ZB, Zheng X, Zhang J, Lin WS, Guo WJ, Li C, et al. The stability of metal halide perovskite nanocrystals—A key issue for the application on quantum-dot-based micro light-emitting diodes display. Nanomaterials. 2020;**10**(7):1375. DOI: 10.3390/nano10071375

[57] Zhang L, Ju MG, Liang WZ. The effect of moisture on the structures and properties of lead halide perovskites: A first-principles theoretical investigation. Physical Chemistry Chemical Physics. 2016;**18**(33):23174-23183. DOI: 10.1039/c6cp01994c

[58] Lin Y, Zheng X, Shangguan Z, Chen G, Huang W, Guo W, et al. All-inorganic encapsulation for remarkably stable cesium lead halide perovskite nanocrystals: Toward full-color display applications. Journal of Materials Chemistry C. 2021;**9**(36):12303-12313. DOI: 10.1039/D1TC02685B

[59] Liu LG, Deng LG, Huang S, Zhang P, Linnros J, Zhong HZ, et al. Photodegradation of organometal hybrid perovskite nanocrystals: Clarifying the role of oxygen by single-dot photoluminescence. Journal of Physical Chemistry Letters. 2019;**10**(4):864-869. DOI: 10.1021/acs.jpclett.9b00143

[60] Zhang Y, Cai GF, Fang Y, Han GJ. Relaying transmission for multiresolution M-DCSK modulation in multi-relay networks. In: 4th International Conference on Communication and Information Systems (ICCIS); 2019 December 21-23. Wuhan. New York: IEEE; 2019. p. 1-6

[61] Quan LN, Rand BP, Friend RH, Mhaisalkar SG, Lee TW, Sargent EH. Perovskites for next-generation optical sources. Chemical Reviews. 2019;**119**(12):7444-7477. DOI: 10.1021/acs.chemrev.9b00107

[62] Rajagopal A, Yao K, Jen AKY. Toward perovskite solar cell commercialization: A perspective and research roadmap based on interfacial engineering. Advanced Materials. 2018;**30**(32):1800455. DOI: 10.1002/adma.201800455

[63] Wu T, Lin Y, Huang Y-M, Liu M, Singh KJ, Lin W, et al. Highly stable full-color display device with VLC application potential using semipolar micro-LEDs and all-inorganic encapsulated perovskite nanocrystal. Photonics Research. 2021;**9**(11):2132-2143. DOI: 10.1364/PRJ.431095

Chapter 2

Milestone Developments and New Perspectives of Nano/Nanocrystal Light Emitting Diodes

Jyoti Singh, Niteen P. Borane and Rajamouli Boddula

Abstract

Light emitting diode (LED) is a one type of p/n junction semiconductor device which is used in less energy consumption for numerous lighting functions. Because of their high performance and long existence, their eye-catching application is getting increasing numbers in recent times. LEDs are nowadays defined as using the "ultimate light bulb". In a previous couple of years, its efficiency has been multiplied through converting it to nano size. This new light-emitting has a nano-pixel structure and it affords high-resolution performance and the geometry of the pixel is cylindrical or conical form. Due to the fact that the previous few years, a few impurity-doped nanocrystal LEDs are varying a good deal in trend. Its performance is very excessive and consumes a smaller amount of voltage. Its monochromatic behavior and indicator excellent are shown publicly demanded in the market and in this work, it's covered evaluations of the fundamental's standards of LEDs and the specific mixed metallic and nanocrystal shape of emitters. In addition, it covers the upcoming challenges that the current trend is working to resolve to get efficient materials to fulfill the future energy crisis.

Keywords: nanostructure, impurity doped nanocrystal, metal mixed LEDs

1. Introduction

A light-emitting diode is one of the eminent revolutionary devices where several applications are currently running such as optoelectronics, lighting, sensing, and medical applications. Light-emitting materials are gained much attention in recent days due to their capacity to reduce the consumption of global electricity. It affects the usage of fossil fuels since more than 20% of electricity is globally consumed and needs an alternative to existing lighting materials. Early stage expected that the Solid State Lighting (SSL) is a type of lightning in LEDs to saved 50% of the electricity consumption of lighting which can reduce 300 x 10^6 tons of annual carbon emission [1, 2]. In addition, commercially available white LEDs moving towards 20–30% efficiency via exhibits fluorescent emission as lamps in limited applications which can be realized by primary color renders (blue, green, red). However, achieving the white LEDs with high color rendering quality is quite challenging and many research groups are working to reach 70–100% efficiency [3]. There is an alternative to reach maximum efficiency by eliminating blue LEDs where efficiency droop arises. To improve efficiency, eminent research focuses attended on nanocrystal-based materials for optoelectronics. In other ways to develop primary colors emitting materials to achieve efficient white light generation, P. H. Fu et al. developed a GaN-based multi-quantum well light emitting diode using SiO_2 nano-honeycomb arrays, natural lithography, and reactive ion etching. Nano-honey comb light output power was found to be 77.8% while using a 40 mA driving current [3]. H. Perlman et al. discussed a nano light emitting device performance enhancement designed for high-resolution display [4].

Further, lighting applications are increasing interest in the field of doping-free and impurity doping materials usage. It is used for LED/Organic LEDs applications where the required resolution high and easy to implement nano-LEDs. In this area still, work is going on phase change materials between two layers. Another focusing area is InGaN/GaN quantum dots structural and optical implementation, along that focused on nano-pixel matrix [5, 6]. Single nanowire pixels are also reported with a full-color demonstration by Y. Ra et al. [7]. Single chip and multi-color nanowire LED pixels are represented in **Figure 1a** and lighting technology efficiencies are in **Figure 1b**. This nanowire single chip arranged InGaN/GaN LEDs exhibits a turn-on voltage of approx. 2 V with no leakage current. Similarly, nano-pixel matrices prepared with metal ions conjugation into polymers by S. Basak et al. Here, RGB colors are arranged in a hierarchical 3-dimensional way and a layer of high-resolution cross strips printed with Eu and Tb metal ions. It was designed at approximately $100 \times 100\ \mu m^2$ and is ca. $0.64 \mu m^2$ [8].

There several other methods are emphasized to develop the lighting materials to enable high luminous efficiency. Additionally, by deliberately introducing atoms or ions of various impurity-doped elements (such as alkali metals, rare earth, lanthanide impurities, and transition metals) into host lattices or non-stoichiometry-induced self-doping, it is possible to produce diverse impurity-doped nanocrystals with desired properties and functions. Because the self-quenching and reabsorption caused by an enhanced Stokes shift may be eliminated, impurity-doped nanocrystals are significantly less sensitive to thermal, chemical, and photochemical perturbations than ones that are not. More holes (p-type doping) or electrons (n-type doping), in particular, are provided with the use of impurities, improving electrical applications. The doping levels and dopant placements are altered by the synthesis schemes (such as doping agents, reaction parameters, and operation temperatures), which also alter the dopant luminescence and electronic impurities [9, 10].

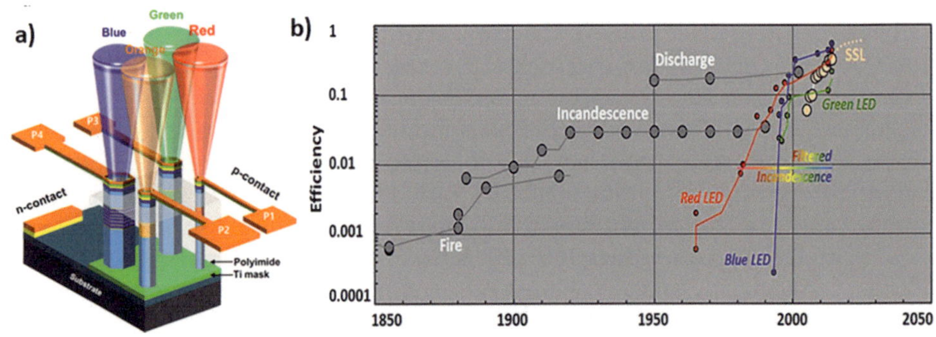

Figure 1.
a) Single chip arranged monolithically integrated multi-color nanowire LED pixels, b) development of the lighting technology efficiencies. Image permission was taken from ref [2, 7].

Recently, interest in impurity-doped nanocrystal light-emitting diodes (LEDs) has increased in both academia and industry due to their great potential to meet the rising need for lighting, display, and signaling technologies. Impedance-doped nanocrystal LEDs have been shown to have a number of advantages over their undoped counterparts, higher brightness, including higher efficiency, longer stability, and lower voltage. In addition to the inherent advantages of nanocrystals, impurity-doped nanocrystals frequently exhibit additional advantages such as enhanced chemical and thermal stability, increased photoluminescence quantum efficiency (PLQY), reduced Auger recombination, customized charge mobility, and impurity-related emission. Because of these benefits, impurity-doped nanocrystals have inspired efforts to satisfy the requirements of several optoelectronic applications [11, 12].

As with many LED emitters, impurity-doped nanocrystals have been intensively studied. Band-edge and impurity-related emissions can frequently be seen in impurity-doped nanocrystal LEDs. Impurity-doped nanocrystal LEDs consequently display three emission behaviors (i.e., LEDs exhibit only host emissions, LEDs show only impurity emissions, and LEDs possess both host and dopant emissions). This is distinct from band-edge emissions exclusively observed in undoped nanocrystal LEDs. Impurity-doped nanocrystal LEDs can also have higher efficiency and brightness compared to their undoped counterparts. For instance, impurity doping was utilized to increase the brightness of PeLEDs by nearly ten times and the external quantum efficiency (EQE) of CQW-LEDs by nine times, respectively. The stability of impurity-doped nanocrystal LEDs may also be higher than that of ones that are undoped. Impurity-doped nanocrystal LEDs, particularly for PeLEDs, CQW-LEDs, and CQD-LEDs, are very promising for upcoming lighting, display, and signaling technologies (such as improved luminance, voltage-increased stability, enhanced efficiency, and lowered power consumption) due to their unique characteristics and amazing benefits [9].

During forecast period 13, which runs from 2021 to 2026, the worldwide OLED display market is expected to grow significantly. The market is expanding steadily in 2020, and due to major players' increasing adoption of strategies, the industry is anticipated to increase during the expected time frame. Considering the influence on the global OLED display market it impacted from both global and regional perspectives. The main key players in OLEDs are Europe, North America, Japan, and China, as per the report put emphasis on the analysis of the market corresponding response policy in different regions [13]. An OLED is a carbon-based light-emitting diode that produces light when current passes through the conductors, the anode and the cathode. With OLED display technology, it delivers superior image quality compared to other display technologies such as liquid crystal displays (LCDs) and LEDs. According to a CAGR of 2021–2026, the OLED display market-generated million of USD in revenue in 2016, million of USD in 2021, and millions of USD in 2026, respectively [14].

2. Development of LEDs

2.1 Milestone inventions

Because of their extraordinary benefits, such as excellent brightness, high efficiency, impressive power consumption, extended lifetime, and low voltage, nanocrystal light-emitting diodes offer enormous promise in signaling, lighting, and display applications [15–17]. In 1994, Alivisatos et al. published the first nanocrystal LED with a maximum EQE of 0.01% using CdSe colloidal quantum dots (CQDs) [18]. A

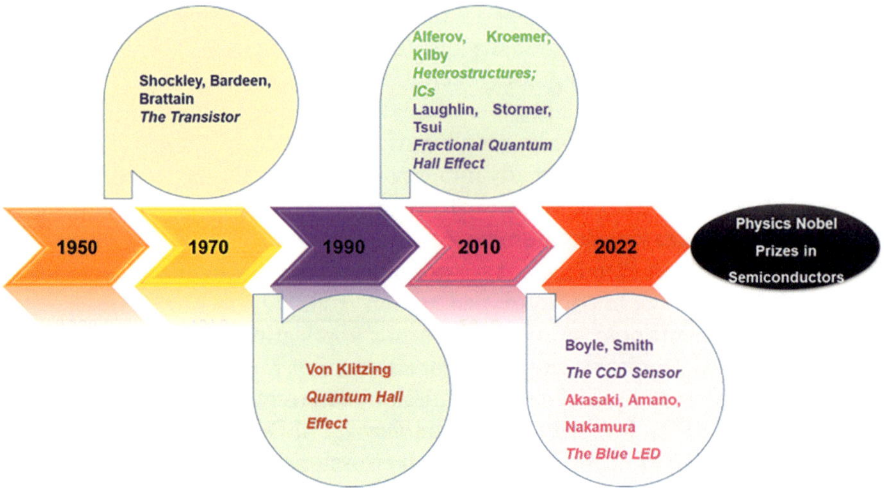

Figure 2.
Physics Noble prizes in semiconductors.

widely used technique to give nanocrystals a variety of, optical, catalytic, new electrical, transporting, and magnetic capabilities is impurity doping [19, 20]. Before the 20th century, four types of semiconductor technology generations are stepping stones for the semiconductor LEDs and the development of the semiconductor technology duration wise demonstrated in **Figure 2** [21, 22].

2.2 Past technology

The most prominent SSL sources are blue GaN/InGaN LEDs with yttrium aluminum garnet (YAG) phosphor. Correlated color temperatures (CCT) were found to be 4000–8000 K for white light generation by mixing this broad yellowish emission of YAG phosphor and blue LED. The color rendering indices (CRI) are recognized to be below 80. These results are somewhat recognizable however, indoor illumination applications required a CCT of less than 4000 K and a CRI value of more than 80. To overcome these limitations, several analyses are invented to improve efficiency. Among them, recently nanocrystal-based materials have expanded the possibility of new milestones. In addition, these materials are capable of tunable and narrow emission visible spectral range. Further, the small overlap between the absorption and emission spectra indicates fewer strokes shift values. Few of the reports stated CRI above 80 by using the dual hybridization of polymers and nanocrystals on LEDs [23–26]. The tunability was achieved by using layer by layer combination of CdSe/ZnS and nanocrystals packed on polyfluorene.

2.3 Knowledge gap and current trend

All the LEDs mentioned above are being used nowadays to make the hundredth-lumen-per-lamp, in which Green, Blue, Red, and White LEDs are used, and red LEDs are the most effective. For this, the LEDs have been prepared with solid-phase materials. The most common application of these LEDs is being used to make displays. Red-green-blue (RGB) LEDs or white LEDs are used to backlight ultra-large video displays or LCDs. Initially, this backlighting is used for mobile displays later on it is used for computer and television displays. The most common use of LED is now in LCD backlighting. This backlighting usually requires more color, but not much, white point-source LEDs are employed for this. However, in the long run, the future of LED backlighting is not certain as LCD has no features. The main reason for this is organic LEDs (OLEDs) have made tremendous progress recently it has greater light emitting efficiency and long life than LEDs and the second reason is LEDs fabrication cost is too high for OLEDs [27].

Devices made of LEDs need more research in the past few decades because of the great advancement in it. In particular, we can envision a time when solid-state lighting is both intelligent and incredibly efficient. This type of intelligent, ultra-efficient solid-state lighting would enable: very high (and gt; 150%) and quot; Effective and quot; light output and usage efficiencies; a variety of new system applications (including feature-rich lighting, and integrated lights/displays), which need to be expanded to various areas of human welfare, as well as technology. In addition, nanotechnology incorporation into the LEDs era is one of the boosting points to develop efficient LEDs. It also noted hybrid light emitting materials tune the emission behavior such as multiple nanocrystals pumped by blue nitride emitters expressed tunability properties. Another preference is to use ultraviolet emitting LEDs by using the nanocrystal thinner films which overcome the blue pumping. Here, the photoemission arises from nanocrystals and direct not depend on the LED platform. It leads to fetching the property of tunability of the color. UV LEDs promising features can have the possibility to reach higher optical levels.

3. CIE, CCT and color rendering index parameters

Nano-crystallized LED materials are characterized by CIE parameters; CIE is "Commission International de l' Eclairage" this method provides basic information about a specific range of light exposed by the nanomaterials (LEDs). The chromaticity variations of light sources for lighting have been offered by the Macadam ellipses' (Macadam, 1942). The CIE color area is purpose monochromatic primary colors with wavelengths of 435.8 nm for blue and 546.1 nm for green, 700 nm for red [28, 29]. **Figure 3** shows the chromaticity graph representing the x, and y-axis.

H. Orucu, et al. have synthesized tridoped Yb/Er/Tm (III) metal-metal ion gadolinium gallium garnet nanocrystal by Solgel Pechini method and its size was determined by X-ray diffraction [30]. It is nearly about 26–56 nm and its CIE values $Gd_3Ga_5O_{12}$: 1% Er^{3+}/1.5%Tm^{3+} 2%Yb^{3+} observed in white space = 0.03244, y = 0.3297 at ordinary temperature under 975 nm IR excitations spectrum. P. Fu and co-workers' nano honeycomb light-emitting diodes (GaN) have 77.8% efficiency [31]. Further G.M.Wu and others increased the extraction effectiveness of GaN/GaN by including small-size photonic inside the LED [32]. The lighting peak's integrated area expanded by 75%, its intensity by 91%, and its integrated area by 106%. T. Xiang, et al. prepared perovskite crystal of 12-crown-4 ether complexes of CAIG PCs, its external QE is 16% [33].

Recently modified high-power LED was prepared using in order to get a higher luminous flux requirement but it has been found that 75–90% of the input power is dissolute as heat so proper heat management is needed for the optical performance of LED for long time temperatures dropped the LED efficiency 1% when the temperature is increased 1°c. K. Yen Yong, et al. synthesized graphene-coated nanoplatelets it reduced the excess temperature and noteworthy improvements in LED cooling [34].

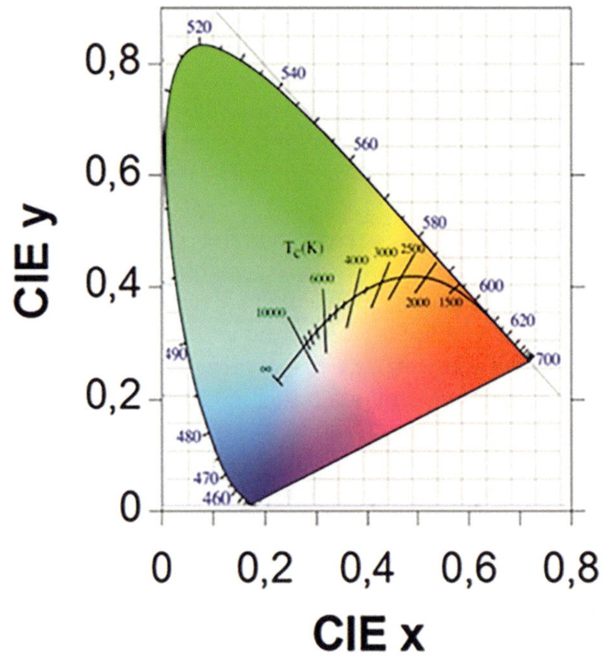

Figure 3.
Chromaticity diagram as CIE.

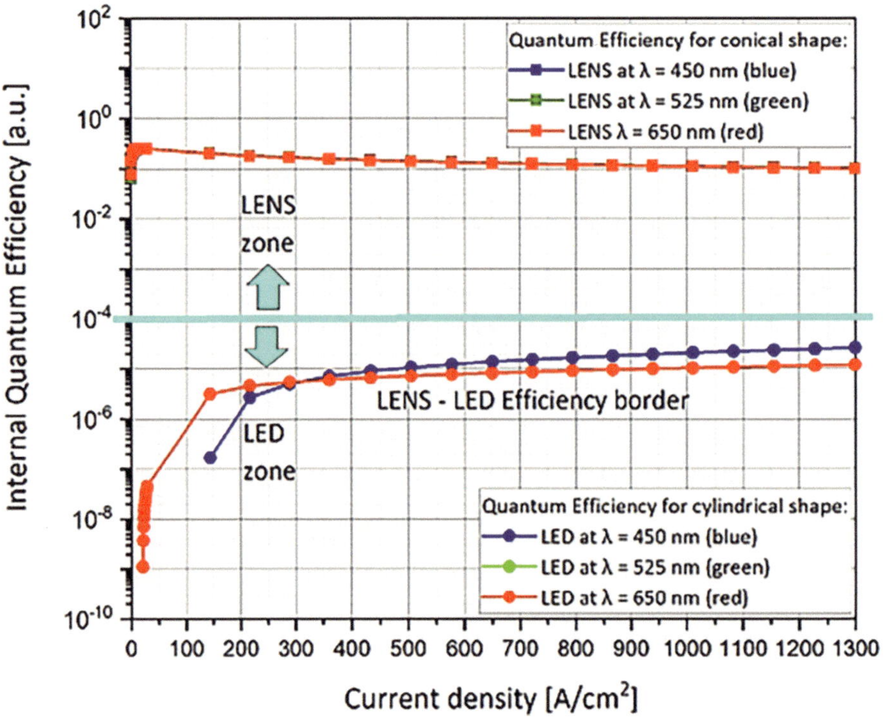

Figure 4.
Nano-pixel plots that show competence graph for cylindrical devices vs. conical devices. Image permission was taken from ref [4].

Microscale light-emitting diodes without sidewall emission have been created by Xinpei Hu and colleagues [35]. In comparison to traditional vertical μLED pixels, the inclined μLED pixel with ODR can boost its overall light extraction efficiency by 2.24 times while reducing sidewall emission by 99.6%. H. Perlman, et al. discussed emission spectra of GaN layer with different color ranges [4]. **Figure 4** indicate different overlapping curve of green, blue and red lines with ranges of color spectra.

4. Standards and design of nanocrystals LEDs

Long-lasting and brighter nanocrystal characteristics for LEDs. The actual light-emitting LED chip is the main component of an LED diode, and it is enclosed in a clear protective domed shell or lens. While a large portion of the light generated by the chip simply travels through the shell, part of it is reflected inward to get around these issues in LEDs.

Nanocrystals are attracting noteworthy consideration for nano-electrical applications for the growth of new non-volatile, high-density appliances. Nanocrystal show many opto-chemical properties, but control of shape and size has been an interesting topic for the advanced growth of nanocrystals. Nano-crystal LEDs also known as light-emitting nano-pixel structured devices have high-resolution displays [4, 36]. It is a unit cell of many more composite structures of LENA (Light Emitting Nano-pixel Array). Nano LEDs improve the feasibility analysis with more clarity vision's these devices are made of two portions, 1st one is nano cone LEDs (**Figure 5**) and 2nd portion is a parabolic concentrator out of these two parts 1st one is very important in order for the reduced the total internal reflection and improve light extraction efficiency. **Table 1** shows a comparative analysis of nano-pixel techniques and an outline of the designed device. When stimulated with a 365 nm LED, some devices exhibit white emissions with excellent CIE coordinates (0.35, 0.31) and a very high color rendering index of 93.

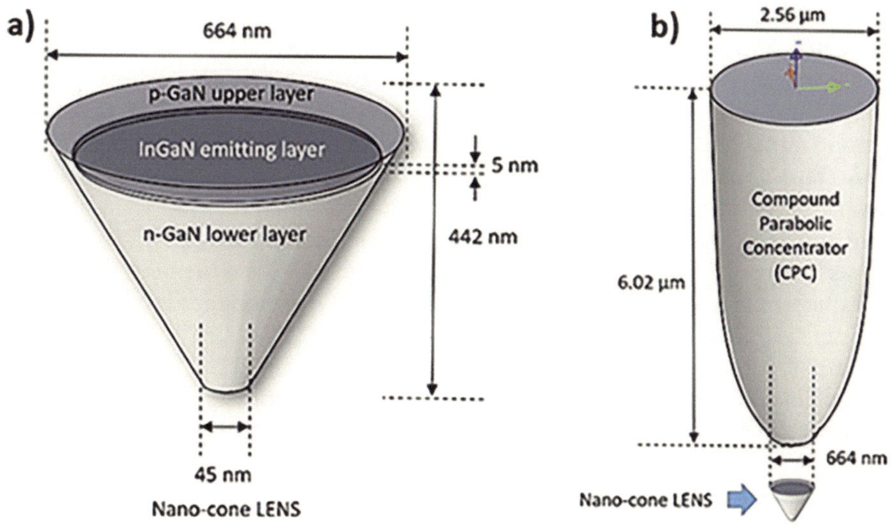

Figure 5.
Represented three-dimension structures of device (a) LENS and (b) CPC. Image permission was taken from ref [4].

Dimensions of pixels	Methods'	Emission	Material	Ref.
300 nm × 300 nm	Phase change material Amorphous to a crystalline state	Non-emitting pixel Not colored. Require backlight illumination	Ge-Sb-Te + Indium Tin Oxide (ITO) electrode	[36]
500 nm × 1000 nm	Dot in nanowire light emitting diode Varying nanowire diameter modulates wavelength emission	Self-emitting pixel RGB	InGaN/GaN	[7]
640 nm × 640 nm	Organic LED (OLED) Hierarchical multi-color nano-pixel matrices	Self-emitting pixel Multicolour	Ligand Polymer + Layer of Eu and Tb ions.	[8]
800 nm diameter	Tunable wavelength InGaN/GaN Nano-ring LEDs via Nano-sphere lithography	Self-emitting pixel RGB	InGaN/GaN	[37]
664 nm diameter	Sub-micron dimension conical advanced shape nano-led	Self-emitting pixel monochromatic RGB option	P- GaN/InGaN/ NGaN	[4]

Table 1.
Comparative analysis of nano-pixel techniques and their specifications.

5. Nanocrystals for LEDs

5.1 White LEDs

Generally, traditionally structured white light-emitting diodes (w-LEDs) be contingent on 3-way combination approaches depending on blue LEDs, yellow-red phosphors and organic resources combinations. However, it frequently suffers from difficult predicaments, including the particularly high fabrication price of traditional yellow and pink phosphors and relatively low robustness due to peripherally natural encapsulating materials. These obstacles have significantly disadvantaged the similarly commercially increasing white light emitting diode (w-LEDs). For this mono-vanadate phosphorus films that had been immediately fabricated on an organic framework at the normal room, temperatures were investigated for w-LEDs. Nanostructures have small nano size compared to other materials like the wavelength of light and are particularly perfect to decorate light interactions [37]. There are so many techniques available for defining the nanostructures' and the morphology of metallic surfaces like surface plasmon polariton (SPP) resonance is of unique interest in this particular area [38]. In 2015 a unique excessive-overall performance phosphors-loose w-LEDs, which is designed by nano-single crystal $Ba_2V_2O_7$ or $Sr_2V_2O_7$ quantum dots (BVQD or SVQD) at once growing on common quartz substrates throughout the deposition of polymer (PAD). As compared to the metavanadates-primarily based phosphor thin layers, the excellent quality of BVQD or SVQD affords a broader band spectrum at 400-700 nm for white LEDs with a 95% quantum yield.

Greater prominently, for homogenous nano-mono crystals di-vanadates quantum dots had been effectively full-grown on not uncommon quartz substrates. Relying on the specific position of polymer-attached metals brings a remarkable leap forward within the subject of heteroepitaxial growth [39]. In 2016 metallic nanostructures assisting plasmonic resonances is a thrilling alternative to this method because of their strong connection between light and matter be counted interplay, which simplifies regulate of light emission without necessitating outside secondary optical fragments. Colloidal semiconductor nanoplatelets (NPLs), a new class of semiconductor nanocrystals with particular structural and electrical characteristics resulting from their flat ring design, have received attention in 2018. It has been shown that Type II NPLs have enormous promise for optoelectronic devices like solar cells and lasers. Here, type II NPL-based nanocrystal light-emitting diodes have been created. These type II NPLs (CdSe/CdSe 0.8 Te 0.2 core/crown) are in use, and their photoluminescence quantum yield is near 85%. Due to their easily adjustable band gaps, strong photoluminescence quantum yield, pure color emission, and cheap cost, inorganic cesium halide perovskite nanocrystals have received a lot of attention in the same year for application in optoelectronic applications. Cesium lead bromide ($CsPbBr_3$) is an inorganic perovskite in which all the bonded atoms are inorganic in nature. However, the structural and optical features of $CsPbBr_3$ nanocrystals deteriorate when they are converted from colloidal solutions to solid thin films, which causes problems with device functioning [40]. This is because organic surfactants are unavoidably used throughout the synthesis. At the same time, the field of optoelectronics has seen a complete revolution owing to hybrid organic-inorganic metal halide perovskites, with exponential growth in efficiency seen for both photovoltaic and light emission applications [41].

The amount of energy consumed worldwide has been rising throughout time. The need for research into sustainable and renewable energy sources is driven by the finite availability of fossil fuels. One of the most promising research to fulfill the rising energy needs of future generations without harming the climate is the conversion of sunlight into electricity. Direct conversion of photon energy into electricity is made possible by solar cell technology, which is environmentally beneficial and renewable [42]. Unfortunately, scientists have not yet been successful in developing photovoltaic devices that are extremely efficient, affordable, and scalable. A unique type of semiconductors has recently developed, the organic-inorganic perovskite with an ABX_3 structure, where X is Cl, Br, or I, and A is cesium (Cs), methyl ammonium (MA), or formamidinium (FA). Perovskites can be treated using a variety of methods, including spray coating and a nozzle that diffuses tiny liquid droplets on substrates for the perovskite layer [43].

Colloidal nanocrystals of organic-inorganic hybrid perovskites (OIHPs), which have exceptional photophysical properties, were a new class of solid-state lighting materials in 2019. An in-depth study has been put into developing high-performance light-emitting diodes based on these materials, including interface engineering, which is crucial for balancing the injection of electrons and holes in gadgets. The effective perovskite nanocrystal LEDs are based on the high electron density 9,10-bis(N-benzimidazolyl) anthracene (BBIA), a novel electron transport material (ETM). The anthracene-based compounds might present fresh study directions for perovskite LED interface engineering [44]. In order to meet the individualized requirements of cutting-edge applications like mobile phones, wearable watches, virtual/augmented reality, micro-projectors, and ultra-high-definition TVs, micro-light-emitting diodes (m-LEDs) are regarded as the foundation of next-generation display technology in 2020. However, due to the limited absorption cross-section, standard phosphor color conversion cannot provide enough brightness and yield to enable high-resolution screens as LED chip sizes decrease to below 20 m. Due to their exceptional photoluminescence, narrow bandwidth emission, color tunability, high quantum yield, and nanoscale size, quantum dot (QD) materials are anticipated to arise and fill this gap, offering a potent full-color solution for -LED displays [45]. White light-emitting diodes (WLEDs) made of silicon nanocrystals (SiNCs) were announced as taking the place of gallium nitride (GaN)-based products that currently dominate the solid-state general lighting market by the end of 2021. Today's high-power IGBT (insulated gate bipolar transistor) devices, wireless and fiber communications, space and compound solar cells, and improved shape memory alloys all depend on Ga in an irreplaceable way. The observed SiNCs have a photoluminescence quantum yield (PLQY) of 11.4% [46].

Jonathan D. Gosnell, et al. designed a new monodispersed nano-size CdSe phosphors excited LEDs to overcome the earlier problem related to LEDs emissions [47]. It gives a large spectrum of 420–710 nm and a small value of quantum yield (10%). Due to its high potential application of nano-size receiving much concentration. White light emitting diodes have various beneficial qualities like high storage capacity and long life and low voltage uses etc. therefore these materials have very high demand [48]. Earlier prepared semiconductor hybrid materials are used for white light emitting diodes like CdSe, CdSe/ZnS, and CdSe/CdS/ZnS but it is very costly as well as associated high energy power consumptions. To overcome these problems Jie Chen, et al. [16] synthesized new silicon-based carbon dots/nanocrystals for LEDs. Carbon nanodots have fascinating properties like it as thermal stability and biocompatibility, eco-friendly materials. Further, it was fabricated by APTES by hydrothermal method. Yanqin Li and coworkers created three CdSe/ZnS QDs of various sizes and blended them in a CBP in an organic matrix to create optically stable ternary nanocrystal composites [49]. A stirring procedure of MS and with $CsPbBr_3$ NCs blended in toluene solution was used by Xiaoxuan Di and colleagues to create $CsPbBr_3$NCs mesoporous silica based light emitting, and the resulting NCs-MS based WLED realizes an easy chromaticity tuning [50].

Combinations of CdSe/ZnS core-shell nanocrystals hybridized on InGaN/GaN LEDs for high color rendering index are used to provide warm white light with these color-converting nanocrystal emitter combinations. Three different sets of proof-of-concept devices are created to provide high-quality warm white light, with the following specifications: 1 (x, y) = 0.37, 0.30, LE = 303 lm/W, CRI = 79.6, and CCT = 1982 K; 2 (x, y) = 0.38, 0.31, LE = 323 lm/W, CRI = 81.0, and 3190 K; and 3 (x, y) = 0.37, 0. CdSe nanocrystal for LEDs was synthesized by Michael A, et al. having excellent color properties and it is defined by CIE color coordinates (0.333, 0.333) and color temp 5461–6007 K with a higher index 96.6 [51]. A simple two-step solution synthesis method was used to create all-inorganic $CsPbBr_3$ perovskite nanocrystals with a mean size of about 300 nm. Although a $CsPb_2Br_5$ impure phase is also present in the finished product, high-resolution transmission electron microscopy and X-ray diffraction characterizations reveal the $CsPbBr_3$ nanoparticles are single crystals with good crystallinity as synthesized [52]. Using an electrospray (e-spray) deposition of a silazane (SZ) oligomer-decorated PeNC solution, He Cheng Yoon and co-authors synthesized $CsPbBr_3$ perovskite nanocrystal (PeNC)-embedded inorganic polymer film by encapsulating the PeNC in a Si N/Si O-based matrix [53]. The LE and CE rates of the white LED were 71.0 lm/W and 50.8%, respectively, at a corresponding color temperature of 9334 K, with only an 8% decline in the LE during long-term operation of 100 h. This indicates that the photophysical properties of these molecules are superior compared to earlier ones. The mechanism of conversion of colors of LEDs for White LEDs is produced using blue LEDs, as seen in **Figure 6** [54].

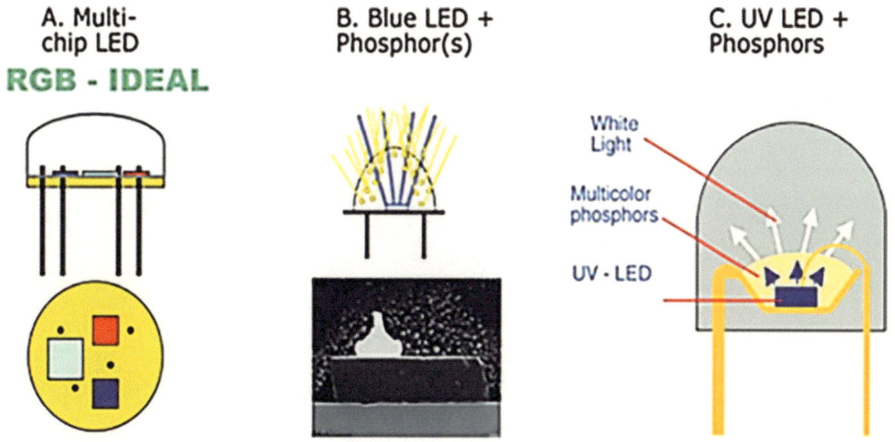

Figure 6.
Designed mechanism of conversion of colors of LEDs. Image permission was taken from ref [54].

5.2 Blue LEDs

The light-emitting diode is a type of semiconductor which play an important role in low lightening power consumption therefore there are a lot of trends of LEDs in the market, but blue LEDs are also creating a lot of attention. In the 1990s Prof Isamu Akasaki and Hiroshi and Shuji Nakamura prepared the first blue LEDs they also described the brightness of LEDs by (InGaN) systems due to less consumption of power blue LEDs have various applications. With advances in phosphor, blue LED efficiency rises from 130 to 140 lm/W to 200–210 lm/W between 2014 and 2018. Therefore, using blue LEDs will significantly reduce lighting costs and benefit the environment. In developing nations, blue light diodes are employed in electronics, solar-powered applications, and indoor and outdoor lighting. Steady enhancement in mug up with accessibility of all colors unexpectedly raised up LEDs like candidates for monochrome application [54]. Here we discussed year-wise progress in LEDs technology. **Figure** 7 indicates year-wise growth in LEDs technology.

Wenjie Xu, et al. have designed new in-situ perovskite blue nanocrystal diodes by the antisolvent process. The LEDs describe all means of EL at 465 nm and EQE values of 2.4% and 2.5 cd A^{-1} separately, by directly adding ligands to the perovskite precursor solution. Nanocrystal halides $(CsPbBr/Cl)_3$ nanocrystals (NCs) play a significant role in blue light emitting diodes [55], the use of dipolar ions of the NCs used in two hole-transporting layers to overcome earlier issues.

Copper-containing tertiary (I-III-VI) chalcogenide nanocrystals (NCs) are prepared by Eric C. Hansen and co-workers by the combinations of lead, and cadmium [56]. The photophysical qualities of these compounds found a PL maximum of 450 nm. En-Ping Yao and others designed inorganic M-X (metal halide) composite perovskite nanocrystal universal light emitting diodes' perovskite nanocrystals (NCs) [16]. Recently so many metals halide perovskite structure blue diodes are synthesized by univalent organic cation. Due to their superior optoelectronic characteristics, such as their narrow emission line widths, high photoluminescence quantum yield (PLQY), tuneable emission wavelength, and high color purity, perovskite-based light-emitting diodes (PLEDs) have become a promising alternative. For green, red, and near-infrared PLEDs with a high external quantum efficiency of about 20%, notable

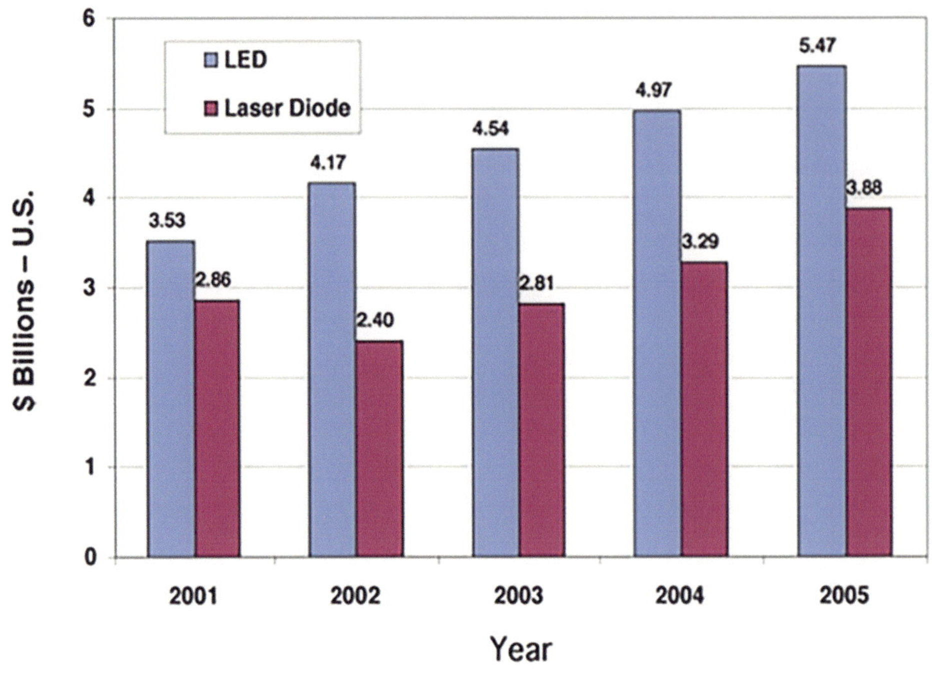

Figure 7.
Wide-reaching application of LEDs in the market. Image permission was taken from ref [54].

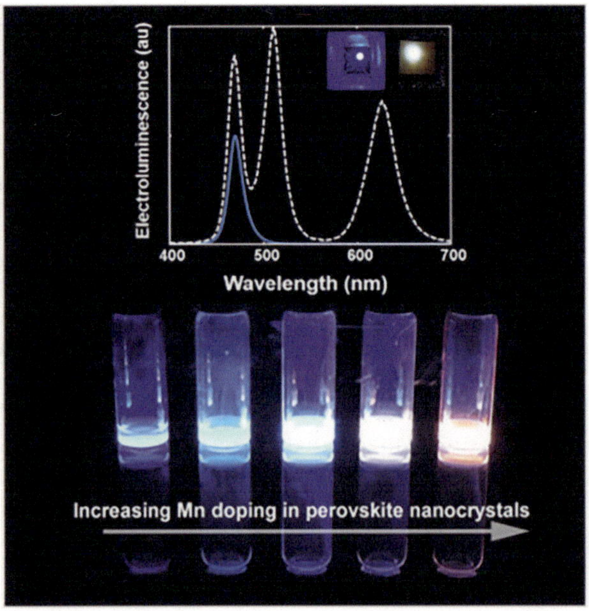

Figure 8.
Effect of Mn addition on electroluminescence spectra. Image permission was taken from ref [58].

progress has been made during the past few years. However, a number of technical challenges, such as subpar film quality, an inefficient device structure, a higher trap density, and others have restricted the development of blue PLEDs [57]. New perovskite structures of Mn-based blue LEDs were created by Shaocong Hou et al., and a small quantity of Mn was added to improve the device's lighting [58]. Blue LEDs with extremely high efficiency and brightness are produced through mn doping. The range of produced compounds' electroluminescence is shown in **Figure 8**.

5.3 Green LEDs

Organometallic compounds were used by Zhi-Kuang Tan and colleagues to create solid-state organic light emitting [57]. Using DDAX as the lone ligand in the hot-injection synthesis, Y. Shynkarenko and co-workers developed long-chain metal halide LEDs with quaternary ammonium-capped CsPb $(Br_{1-x}C_y)_3$ NCs. Emitting a green light, which has C12-hydrocarbon chains with an EQE of 9.8% and a brightness of 34,700 Cd/m^2, produces the best results [59]. Recently lead halide colloidal nanocrystal are used as an extraordinary candidate for LEDs because it has excellent optoelectronic quality, along with very high PL yields, and small size with wide color tuneable properties. Young-Hoon and others proved highly organized light-emitting diodes constructed by the colloidal nanocrystal perovskite by using a multifunctional buffer hole layer (Buf-HIL), This Buf-HIL coupled with poly(3,4-ethylene dioxythiophene)/poly(styrene sulfonate) and per-fluorinated ionomer with these approaches they achieved excellent yields of photoluminescent quantum yield (60.5%) [60]. Normally light emitting diodes are a combination of the organic framework with inorganic metal halide, Perovskite NCs with dimensions bigger than the DB can be created using technique 10 by using a multipurpose buffer hole injection layer and (ii) (Buf-HIL) **Figure 9**.

Jun Xing, co-author of organic metal framework associated metal halide LEDs and provided a systematic route of a chain of colloidal halide perovskite preparation of $CH_3NH_3PbX_3$ non-crystalline nanoparticles and this shows maximum luminous efficiency of 11.49 Cd/A of 3.8% EQE values and these values higher than previously reported colloidal quantum dot-based LEDs [61]. **Figure 10**, discussed the structure and analysis data of $CH_3NH_3PbX_3$ non-crystalline nanoparticles. Bright and reliable light-emitting diodes were created by Hsinhan Tsai and colleagues using perovskite nanocrystals stabilized in metal-organic frameworks (PeMOF) [62]. Cs-PeMOF is used to further improve the compound's stability because the modification device has the property of thermal stability. Internal quantum efficiency was raised by 25% in the NVPCs reported by Yu-Lin Tsai and others, and droop behavior was decreased from 37.4% to 25.9% [63]. With a driving current of 350 mA, an increase in light output power of up to 151% is possible. Researchers have also created layers of ZnO nanocrystals capable of electron injection in blue, red, and green nanocrystals. The performance of green LEDs is lower than that of their red and blue relatives, which are much more obtrusive and need more work [64].

5.4 Red LEDs

The color efficiency of LEDs depends on the size of the device blue LEDs have already completed all these demands. It was found that green LEDs has some draw-back mainly suffering from an emission peak frequency which is very high and also some color purity issue [62, 65]. LEDs are semiconductors and in the case of semiconductor emissions, colors can be altered by changing the size of the crystal. In the market white, blue and green LEDs are more popular but as the demand for Red LEDs is increasing, so scientists are making Red LEDs by using different materials.

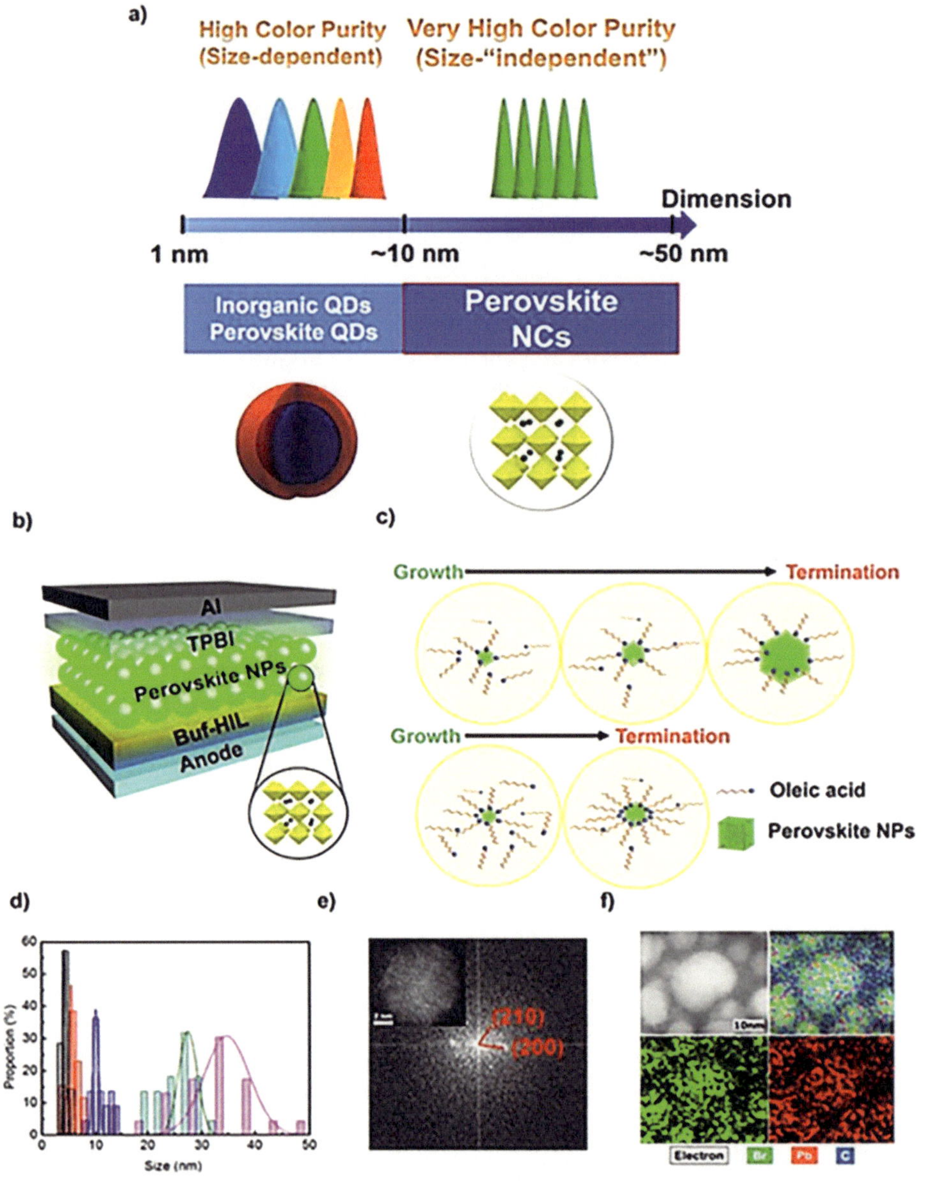

Figure 9.
(a) Distribution of inorganic quantum dots and structure of nanocrystal, (b) device fabrication ways (c) structure of nanocrystal perovskites, (d) histogram of nanocrystals, (c) FTIR and TEM analysis (f) elemental mapping of perovskite NPs. Image permission was taken from ref [60].

Parth Vashishtha and et al. have investigated that mixed halide nanocrystal shows excellent color quality's for the new family of nanocrystal light-emitting diodes. Prepared compounds produce stable red emissions in the red region (620-650 nm) [66]. Mixed halide CsPbBr, X, (X = I or Cl) peNC organic LEDs using peNC emitters with photoluminescence. **Figure 11** illustrated the construction of QD-OLEDs from $CsPbX_3$ nanocrystals (X = Cl, Br or I) grown via a colloidal synthesis procedure. Red NCs (nanocrystals are a very significant component of perovskite light emitting devices but their stability is very low because of the stacked structure so some modification is

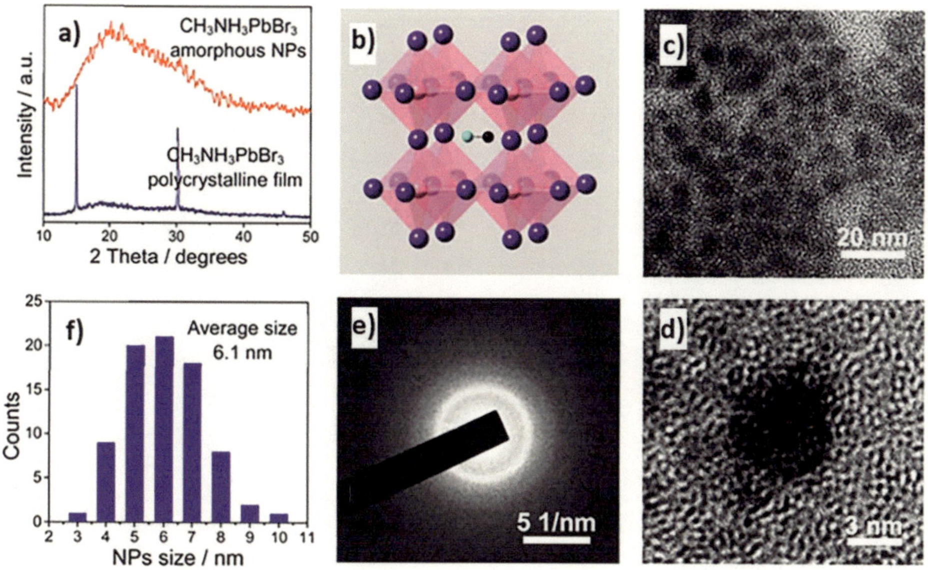

Figure 10.
a) XRD patterns of amorphous $CH_3NH_3PbBr_3$ NPs and polycrystalline film, (b) atomic model of cubic phase $CH_3NH_3PbBr_3$. Blue ball, I; pink ball, Pb; black ball, CH_3; green ball, NH_3. (c,d) HRTEM image, (e) SAED pattern, and (f) size distribution of $CH_3NH_3PbBr_3$ NPs. Image permission was taken from ref [61].

necessary. Some researchers modified its morphology and efficiency by using transition metal therefore luminescent red perovskite $CsPbBrI_2$ NCs were achieved through Cu substitution and halide rich passivation strategy [67]. With a higher photoluminescence quantum yield (up to 94.8%) and enhanced stability, Cu^{2+}-substituted $CsPbBrI_2$ NCs have demonstrated exceptional performance. In order to create high-performance red perovskite LEDs, the very stable and luminous Cu^{2+}-substituted $CsPbBrI_2$ NCs can work effectively as light emitters. Using an epitaxial solution growth technique, Jibin Zhang and colleagues created new, very stable, red-emitting $CsPbBrI_2$/PbSe heterojunction nanocrystals (h-NCs), where each $CsPbBrI_2$ NC was covered in PbSe in the $CsPbBrI_2$/PbSe heterodimers [66, 68].

The purification of the chemicals to create extremely luminous and stable $CsPbCl_{3x}Br_x$ and $CsPbI_3$ PNCs was accomplished using various crystallization techniques of PbX_2 (X = Cl or I), according to Chae Hyun Lee and the co-author [69]. By using a hydrothermal (Hyd) procedure to enhance the quality of the $PbCl_2$ as-prepared, blue-emitting PNCs are given effective ligand surface passivation, a maximum photoluminescence quantum yield (PLQY) of around 88%, and increased photocatalytic activity to oxidize benzyl alcohol, producing 40%. The creation of red-emissive PNCs with a PLQY of up to 100% was then seen as a result of the heated recrystallization of PbI_2 prior to (Hyd) treatment. $CsPbBr_xI_{3-x}$ LEDs based on nano-sized - $CsPbBr_xI_{3-x}$ crystallites have recently been manufactured primarily by the traditional colloidal technique, which includes a laborious procedure of nanocrystal synthesis, purification, ligand or anion, etc. $CsPbBr_xI_{3-x}$ LEDs have only been able to turn on at high turn-on voltages (>2.7) while using the commonly used traditional LED device structure. Additionally, this mix-halide system may experience significant spectra shift under bias. This study describes the preparation of $CsPbBr_xI_{3-x}$ thin films with nano-sized crystallites using a one-step spin-coating method that incorporates several ammonium ligands. The growth of $CsPbBr_xI_{3-x}$ nanograins is restricted by

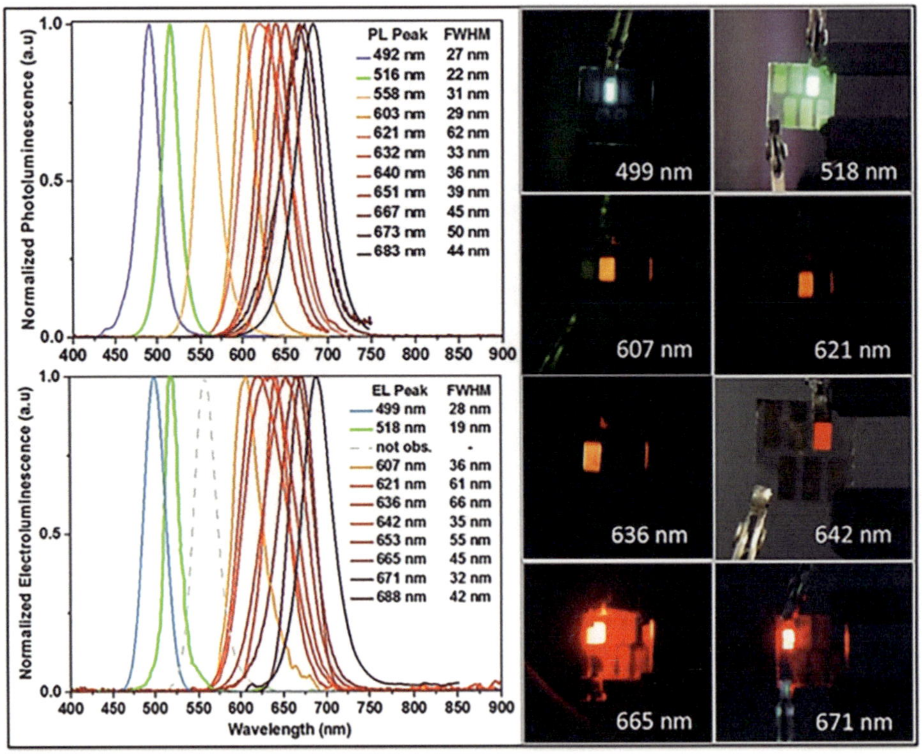

Figure 11.
Normalized photoluminescence spectra (top) taken at 400 nm excitation of a solution of $CsPbBr_{3-x}I_x$, for varying values of x from 0 (green) to 2.75 (deep red). Blue NCs consist of $CsPbBr_{3-x}Cl_x$ (x = 0.75). Normalized electroluminescence spectra (bottom) from ITO/PEDOT: PSS/pTPD/peNC/TPBi/Al LEDs for these same peNCs. Image permission was taken from ref [62].

the various ammonium ligands. Such quantum confinement-benefited $CsPbBr_xI_{3-x}$ thin films are advantageous. The corresponding $CsPbBr_xI_{3-x}$ LEDs emit pure-red color (CIE) coordinates of (0.709, 0.290), (0.711, 0.289), etc., which represent the highest color-purity for reported pure-red perovskite [70]. They use a conventional LED structure of indium-doped tin oxide (ITO) /poly (3,4 ethylene dioxythiophene) poly(styrene).

6. Future perspectives

Here, we faced the common characteristics of all LEDs and also concentrated on the issues relating to the color characteristics and stability characteristics of various types of perovskite forms of LEDs. Here, we addressed the inherent characteristics of all LEDs, and that was our main objective. The main obstacles to obtaining desired stability and efficiency can be roughly divided into two groups: (a) a careful selection of photonic structures, and (b) an understanding of and control of intrinsic material qualities for optimum performance. Additionally, interfacial layers enable smooth integration, improved illumination, and fall avoidance for devices. We assume these approaches to be in addition changed to improve the nice of LED improvement within the destiny.

7. Conclusions

A nano light emitting device performance enhancement design of high-resolution display which is currently focusing area in lighting technology. The white light emission, blue can exempt however red is a common and essential parameter. Blue accepts by yellow is feasible to reach more efficiency and electroluminescence semiconductor light emitters usage is preferable. Which are better than the phosphors filling in the red-yellow-green gap of lighting. It expects to improve efficiency and it leads to improving productivity. Since a few years ago, nanocrystal LEDs have become very popular because of their appealing photophysical characteristics. We have covered all of the physical characteristics and production methods for white, blue, green, and red LEDs in this review post. In addition, discussed a Noble prize works by duration in terms of semiconductors. The necessity for rigorous research into these material systems and the best device configurations. Since the long run, the future of LED backlighting is not certain as LCD has no features. Additionally, there is a need for significant upgrades. In this section, we also discussed how nanocrystals are introduced and color rendering properties. The design of Nanocrystal LEDs with different dimensions is elaborated to understand the reported design structures which help to design new upcoming architectures of lighting.

Acknowledgements

Authors gratefully acknowledge the financial support by the GUJCOST, Gov. of India (Project No. GUJCOST/2020-2021/2012).

Conflict of interest

There are no conflicts to declare.

Author details

Jyoti Singh, Niteen P. Borane and Rajamouli Boddula*
Tarsadia Institute of Chemical Science, Uka Tarsadia University, Maliba Campus, Bardoli, Gujarat, India

*Address all correspondence to: rajamouliboddula@gmail.com

References

[1] Sandia National Laboratories. Sandia National Laboratories Solid-State Lighting: Better and More Effic. Published online 2008. Available frrom: http://www.sandia.gov

[2] Tsao JY, Han J, Haitz RH, Pattison PM. The blue LED Nobel prize: Historical context, current scientific understanding, human benefit. Annalen der Physik. 2015;**527**(5-6):A53-A61. DOI: 10.1002/andp.201570058

[3] Laboratories SN. Solid-State lighting Research & development at Sandia national laboratories. 2008. Available from: http://www.ssls.sandia.gov

[4] Perlman H, Eisenfeld T, Karsenty A. Performance enhancement and applications review of nano light emitting device (led). Nanomaterials 2021;**11**(1):1-27. DOI: 10.3390/nano11010023

[5] Tsao JY, Crawford MH, Coltrin ME, et al. Toward smart and ultra-efficient solid-state lighting. Advanced Optical Materials. 2014;**2**(9):809-836. DOI: 10.1002/adom.201400131

[6] Hosseini P, Wright CD, Bhaskaran H. An optoelectronic framework enabled by low-dimensional phase-change films. Nature. 2014;**511**(7508):206-211. DOI: 10.1038/nature13487

[7] Ra YH, Wang R, Woo SY, et al. Full-color single nanowire pixels for projection displays. Nano Letters. 2016;**16**(7):4608-4615. DOI: 10.1021/acs.nanolett.6b01929

[8] Basak S, Mohiddon MA, Baumgarten M, Müllen K, Chandrasekar R. Hierarchical multicolor nano-pixel matrices formed by coordinating luminescent metal ions to a conjugated poly(4′-octyl-2′,6′-bispyrazoyl pyridine) film via contact printing. Scientific Reports. 2015;**5**(i):8406. DOI: 10.1038/srep08406

[9] Luo D, Wang L, Qiu Y, Huang R, Liu B. Emergence of impurity-doped nanocrystal light-emitting diodes. Nanomaterials Review. 2020;**10**:1226. DOI: 10.3390/nano10061226

[10] Bryan JD, Gamelin DR. Doped semiconductor nanocrystals: Synthesis, characterization, physical properties, and applications. Progress in Inorganic Chemistry. 2005;**54**:47-126 DOI: 10.1002/0471725560.ch2

[11] Kanemitsu Y. Multiple exciton generation and recombination in carbon nanotubes and nanocrystals. Accounts of Chemical Research. 2013;**46**(6):1358-1366. DOI: 10.1063/pt.4.0563

[12] Kriegel I, Scotognella F, Manna L. lasmonic doped semiconductor nanocrystals: Properties, fabrication, applications and perspectives. Physics Reports. 2017;**674**:1-52. DOI: 10.1016/j.physrep.2017.01.003

[13] Laricchia F. OLED panel production capacity share 2016-2025, by country. 2022. Available from: http://www.statista.com

[14] OLED display market size, COVID-19 impact analysis, regional outlook, application development potential, Price trend, Competitive Market Share & Forecast, 2022 – 2028. Glob Mark insights. Published online 2022:Report ID: GMI3827.

[15] Mondal N, De A, Samanta A. Achieving near-Unity photoluminescence

efficiency for blue-violet-emitting perovskite nanocrystals. ACS Energy Letters. 2019;**4**(1):32-39. DOI: 10.1021/acsenergylett.8b01909

[16] Ge C, Fang Q, Lin H, Hu H. Review on blue perovskite light-emitting diodes: Recent advances and future prospects. Front in Materials. 2021;**8**(March):1-7. DOI: 10.3389/fmats.2021.635025

[17] Cui B, Feng XT, Zhang F, Wang YL, Liu XG, Yang YZ, et al. The use of carbon quantum dots as fluorescent materials in white LEDs. Xinxing Tan Cailiao/New Carbon Materials. 2017;**32**(5):385-401. DOI: 10.1016/S1872-5805(17)60130-6

[18] Tolbert SH, Alivisatos AP. Size dependence of a first order solid-solid phase transition: The wurtzite to rock salt transformation in CdSe nanocrystals. Science (80-). 1994;**265**(5170):373-376. DOI: 10.1126/science.265.5170.373

[19] Kovalenko MV, Protesescu L, Bodnarchuk MI. Properties and potential optoelectronic applications of lead halide perovskite nanocrystals. Science (80-). 2017;**358**(6364):745-750. DOI: 10.1126/science.aam7093

[20] Liang S, Zhang M, Biesold GM, et al. Recent advances in synthesis, properties, and applications of metal halide perovskite nanocrystals/polymer nanocomposites. Advanced Materials. 2021;**33**(50):1-36. DOI: 10.1002/adma.202005888

[21] Mittleman DM, Schoenlein RW, Shiang JJ, Colvin VL, Alivisatos AP, Shank CV. Quantum size dependence of femtosecond electronic dephasing and vibrational dynamics in CdSe nanocrystals. Physical Review B. 1994;**49**(20):14435-14447. DOI: 10.1103/PhysRevB.49.14435

[22] Schlamp MC, Peng X, Alivisatos AP. Improved efficiencies in light emitting diodes made with CdSe(CdS) core/shell type nanocrystals and a semiconducting polymer. Journal of Applied Physics. 1997;**82**(11):5837-5842. DOI: 10.1063/1.366452

[23] Demir HV, Nizamoglu S, Ozel T, Nizamoglu S, Ozel T, Mutlugun E, et al. White light generation tuned by dual hybridization of nanocrystals and conjugated polymers. New Journal of Physics. 2007;**9**(10). DOI: 10.1088/1367-2630/9/10/362

[24] Ali M, Chattopadhyay S, Nag A, Kumar A, Sapra S, Chakraborty S, et al. White-light emission from a blend of CdSeS nanocrystals of different Se: S ratio. Nanotechnology. 2007;**18**(7):1-4. DOI: 10.1088/0957-4484/18/7/075401

[25] Nizamoglu S, Demir HV. Hybrid white light sources based on layer-by-layer assembly of nanocrystals on near-UV emitting diodes. Nanotechnology. 2007;**18**(40):1-4. DOI: 10.1088/0957-4484/18/40/405702

[26] Schubert EF. Light-Emitting Diodes. Cambridge, UK: Cambridge University Press. 2006. Available from: https://archive.org/details/lightemittingdio00schu_0

[27] Chynoweth AG. Charge multiplication phenomena. In: Semicond Semimetals. 1968;**4**:263-325. DOI: 10.1016/S0080-8784(08)60345-2

[28] Rajamouli B, Sivakumar V. Eu(III) complexes for LEDs based on Carbazole- and Fluorene-functionalized Phenanthro-imidazole ancillary ligands: Detailed Photophysical and theoretical study. ChemistrySelect. 2017;**2**(14):4138-4149. DOI: 10.1002/slct.201700266

[29] Westland S. Review of the CIE system of colorimetry and its use in dentistry. Journal of Esthetic and Restorative Dentistry. 2003;**15**:S5-S12. DOI: 10.1111/j.1708-8240.2003.tb00313.x

[30] Örücüa H, Sevcan Tabanli ME, Öztürk Y, Eryürekb G. Bright white light up-conversion luminescence from Yb^{3+}/Er^{3+}/Tm^{3+} tridoped gadolinium gallium garnet nano-crystals for multicolor and white light-emitting diodes. Optical Materials (Amst). 2022;**131**:112613. DOI: 10.1016/j.optmat.2022.112613

[31] Fu PH, Lin GJ, Wang HP, Lai KY, He JH. Enhanced light extraction of light-emitting diodes via nano-honeycomb photonic crystals. Nano Energy. 2014;**8**:78-83. DOI: 10.1016/j.nanoen.2014.05.006

[32] Films TS. Effects of nano-structured photonic crystals on light extraction enhancement of nitride light-emitting diodes. GMWu, CCYen, HWChien, HCLu, TWChang, TENee. 2011;**519**(15):5074-5077. DOI: 10.1016/j.tsf.2011.01.131

[33] Xiang T, Li T, Wang M, Zhang W, Ahmadi M, Wu X, et al. 12-Crown-4 ether assisted in-situ grown perovskite crystals for ambient stable light emitting diodes. Nano Energy. 2022;**95**:107000. DOI: 10.1016/j.nanoen.2022.107000

[34] Yong KY, Chan YK, Von LE, Hung YM. Effective passive phase-change light-emitting diode cooling system using graphene nanoplatelets coatings. Case Studies in Thermal Engineering. 2022;**31**(January):101795. DOI: 10.1016/j.csite.2022.101795

[35] Hu X, Cai J, Liu Y, Zhao M, Chen E, Guo Y, et al. Design of inclined omni-directional reflector for sidewall-emission-free micro-scale light-emitting diodes. Optics and Laser Technology. 2022;**154**:108335. DOI: 10.1016/j.optlastec.2022.108335

[36] Pimpin A, Srituravanich W. Reviews on micro- and nanolithography techniques and their applications. Engineering Journal. 2012;**16**(1):37-55. DOI: 10.4186/ej.2012.16.1.37

[37] Wang X, He H, Gao J, Gao J, Hu H, Tang S, et al. Effects of nanoparticle structural features on the light-matter interactions in nanocermet layers and cermet-based solar absorbers. Journal of Materials. 2021;**7**(5):1103-1111. DOI: 10.1016/j.jmat.2021.01.011

[38] Moon H, Bersin E, Chakraborty C, Lu A, Grosso G, Kong J, et al. Surface plasmon subwavelength optics. ACS Photonics. 2003;**424**:824-830. DOI: 10.1021/acsphotonics.0c00626

[39] Yang Z, Liu J, Chen LH, Zhang L, Pan H, Wu B, et al. A high-performance white-light-emitting-diodes based on nano-single crystal divanadates quantum dots. Scientific Reports. 2015;**5**(1):1-7

[40] Wang L, Liu B, Zhao X, Demir HV, Gu H, Sun H. Solvent-assisted surface engineering for high-performance all-inorganic perovskite nanocrystal light-emitting diodes. ACS Applied Materials & Interfaces. 2018;**10**(23):19828-19835. DOI: 10.1021/acsami.8b06105

[41] Kar S, Jamaludin NF, Yantara N, Mhaisalkar SG, Leong WL. Recent advancements and perspectives on light management and high performance in perovskite light-emitting diodes. Nanophotonics. 2020;**10**(8):2103-2143. DOI: 10.1515/nanoph-2021-0033

[42] Shaikh MRS. A review paper on electricity generation from solar energy. International Journal of Research Applications Science Engineering and Technology. 2017;**V**(IX):1884-1889. DOI: 10.22214/ijraset.2017.9272

[43] Wang R, Mujahid M, Duan Y, Wang ZK, Xue J, Yang Y. A review of perovskites solar cell stability. Advanced Functional Materials. 2019;**29**(47):1-25. DOI: 10.1002/adfm.201808843

[44] Kumar S, Marcato T, Vasylevskyi SI, Jagielski J, Fromm KM, Shih CJ. Efficient perovskite nanocrystal light-emitting diodes using a benzimidazole-substituted anthracene derivative as the electron transport material. Journal of Materials Chemistry C. 2019;7(29):8938-8945. DOI: 10.1039/c9tc02352f

[45] Liu Z, Lin CH, Hyun BR, Sher CW, Lv Z, Luo B, et al. Micro-light-emitting diodes with quantum dots in display technology. Light Science Application. 2020;**9**(1):1-23. DOI: 10.1038/s41377-020-0268-1

[46] Zhang YC, Yu ZY, Xue XY, Wang FL, Li S, Dai XY, et al. High brightness silicon nanocrystal white light-emitting diode with luminance of 2060 cd/m^2. Optics Express. 2021;**29**(21):34126. DOI: 10.1364/oe.437737

[47] Schreuder MA, Gosnell JD, Smith NJ, Warnement MR, Weiss SM, Rosenthal SJ. Encapsulated white-light CdSe nanocrystals as nanophosphors for solid-state lighting. Journal of Materials Chemistry. 2008;**18**(9):970-975. DOI: 10.1039/b716803a

[48] Chen J, Liu W, Mao LH, Yin YJ, Wang CF, Chen S. Synthesis of silica-based carbon dot/nanocrystal hybrids toward white LEDs. Journal of Materials Science. 2014;**49**(21):7391-7398. DOI: 10.1007/s10853-014-8413-y

[49] Li Y, Rizzo A, Cingolani R, Gigli G. Bright white-light-emitting device from ternary nanocrystal composites. Advanced Materials. 2006;**18**(19):2545-2548. DOI: 10.1002/adma.200600181

[50] Di X, Shen L, Jiang J, Shen L, Jiang J, He M, et al. Efficient white LEDs with bright green-emitting CsPbBr3 perovskite nanocrystal in mesoporous silica nanoparticles. Journal of Alloys and Compounds. 2017;**729**(Complete):526-532. DOI: 10.1016/j.jallcom.2017.09.213

[51] Schreuder MA, Xiao K, Ivanov IN, Weiss SM, Rosenthal SJ. White light-emitting diodes based on ultrasmall CdSe nanocrystal electroluminescence. Nano Letters. 2010;**10**(2):573-576. DOI: 10.1021/nl903515g

[52] Yue W, Liu Y, Heyu C, Chunyang L, Weizhen L, Haiyang X, et al. White LED based on $CsPbBr_3$ nanocrystal phosphors via a facile two-step solution synthesis route. Materials Research Bulletin. 2018;**104**(February):48-52. DOI: 10.1016/j.materresbull.2018.03.055

[53] Yoon HC, Lee S, Song JK, Yang H, Do YR. Efficient and stable $CsPbBr_3$ quantum-dot powders passivated and encapsulated with a mixed silicon nitride and silicon oxide inorganic polymer matrix. ACS Applied Materials & Interfaces. 2018;**10**(14):11756-11767. DOI: 10.1021/acsami.8b01014

[54] Bergh AA. Blue laser diode (LD) and light emitting diode (LED) applications. Phys Status Solidi Application Research. 2004;**201**(12):2740-2754. DOI: 10.1002/pssa.200405124

[55] Ochsenbein ST, Krieg F, Shynkarenko Y, Rainò G, Kovalenko MV. Engineering color-stable blue light-emitting diodes with Lead halide perovskite nanocrystals. ACS Applied Materials & Interfaces. 2019;**11**(24):21655-21660. DOI: 10.1021/acsami.9b02472

[56] Hansen EC, Liu Y, Utzat H, Bertram SN, Grossman JC, Bawendi MG. Blue light emitting defective nanocrystals composed of earth-abundant elements. Angewandte Chemie International Edition. 2020;**59**(2):860-867. DOI: 10.1002/anie.201911436

[57] Tan ZK, Moghaddam RS, Lai ML, Docampo P, Higler R, Deschler F, et al.

Bright light-emitting diodes based on organometal halide perovskite. Nature Nanotechnology. 2014;**9**(9):687-692. DOI: 10.1038/nnano.2014.149

[58] Hou S, Gangishetty MK, Quan Q, Congreve DN. Efficient blue and white perovskite light-emitting diodes via manganese doping. Joule. 2018;**2**(11):2421-2433. DOI: 10.1016/j.joule.2018.08.005

[59] Smock SR, Chen Y, Rossini AJ, Brutchey RL. The surface chemistry and structure of colloidal Lead halide perovskite nanocrystals. Accounts of Chemical Research. 2021;**54**(3):707-718. DOI: 10.1021/acs.accounts.0c00741

[60] Kim YH, Wolf C, Kim YT, Cho H, Kwon W, Do S, et al. Highly efficient light-emitting diodes of colloidal metal-halide perovskite nanocrystals beyond quantum size. ACS Nano. 2017;**11**(7):6586-6593. DOI: 10.1021/acsnano.6b07617

[61] Xing J, Yan F, Zhao Y, Chen S, Yu H, Zhang Q, et al. High-efficiency light-emitting diodes of Organometal halide perovskite amorphous nanoparticles. ACS Nano. 2016;**10**(7):6623-6630. DOI: 10.1021/acsnano.6b01540

[62] Vashishtha P, Halpert JE. Field-driven ion migration and color instability in red-emitting mixed halide perovskite nanocrystal light-emitting diodes. Chemistry of Materials. 2017;**29**(14):5965-5973. DOI: 10.1021/acs.chemmater.7b01609

[63] Tsai YL, Liu CY, Krishnan C, Lin DW, Chu YC, Chen TP, et al. Bridging the "green gap" of LEDs: Giant light output enhancement and directional control of LEDs via embedded nano-void photonic crystals. Nanoscale. 2016;**8**(2):1192-1199. DOI: 10.1039/c5nr05555e

[64] Janssen RAJ, Stouwdam JW. Red, green, and blue quantum dot LEDs with solution processable ZnO nanocrystal electron injection layers. Journal of Materials Chemistry. 2008;**18**(16):1889-1894. DOI: 10.1039/b800028j

[65] Sciences M. Investigation of InGaN-based red/green micro-light-emitting diodes. Optics Letters. 2021;**46**(8):1912-1915

[66] Shen W, Zhang J, Dong R, Chen Y, Yang L, Chen S, et al. Stable and efficient red perovskite light-emitting diodes based on Ca^{2+}-doped $CsPbI_3$ nanocrystals. Research. 2021;**2021**:1-11. DOI: 10.34133/2021/9829374

[67] Zhang J, Zhang L, Cai P, Xue X, Wang M, Zhang J, et al. Enhancing stability of red perovskite nanocrystals through copper substitution for efficient light-emitting diodes. Nano Energy. 2019;**62**:434-441. DOI: 10.1016/j.nanoen.2019.05.027

[68] Zhang J, Liu X, Jiang P, Chen H, Wang Y, Ma J, et al. Red-emitting $CsPbBrI_2$/PbSe heterojunction nanocrystals with high luminescent efficiency and stability for bright light-emitting diodes. Nano Energy. 2019;**66**:104142. DOI: 10.1016/j.nanoen.2019.104142

[69] Lee CH, Shin YJ, Villanueva-Antolí A, Das Adhikari S, Rodriguez-Pereira J, Macak JM, et al. Efficient and stable blue- and red-emitting perovskite nanocrystals through defect engineering: PbX2Purification. Chemistry of Materials. 2021;**33**(22):8745-8757. DOI: 10.1021/acs.chemmater.1c02772

[70] Jiang M, Hu Z, Ono LK, Qi Y. $CsPbBr_xI_{3-x}$ thin films with multiple ammonium ligands for low turn-on pure-red perovskite light-emitting diodes. Nano Research. 2021;**14**(1):191-197. DOI: 10.1007/s12274-020-3065-5

Section 2

High Power LEDs

Chapter 3

High-Temperature Operating Narrow-Period Terahertz Quantum Cascade Laser Designs

Li Wang and Hideki Hirayama

Abstract

Presently, terahertz quantum cascade lasers still suffer from operations below room temperature, which prohibits extensive applications in terahertz spectra. The past continuous contributions to improving the operating temperatures were by clarifying the main thermal degradation process and proposing different designs with the optical gain demonstrating higher temperature cut-offs. Recent designs have attempted to employ a narrow period length with a simplified and clean state system, and reach renewed operating temperatures above 200 K. This study reveals how historic designs approach such narrow-period designs, discus the limitations within those designs, and show further possible designs for higher operating temperatures.

Keywords: high operating temperature, terahertz, quantum cascade lasers, narrow-period design

1. Introduction

The terahertz (THz) electromagnetic spectrum ($\lambda \sim 600$–30 μm, $\hbar\omega \sim 2$–40 meV) remains the most underdeveloped spectral region, at least largely due to the lack of THz radiation sources with coherent or compact wave features and high output power. The potential for applications in THz spectra is included but is not limited to atmospheric science, astrophysics, biological/medical sciences, security screening and illicit material detection, nondestructive evaluation, ultrafast spectroscopy, and communications technology [1–4]. THz quantum cascade lasers (THz-QCLs) are thought to be the only solid-state THz sources that can offer average optical power levels much greater than milliwatts, which are profitable for imaging applications [5], and continuous-wave (CW) operation for the frequency stability desired in high-resolution spectroscopy techniques.

These laser sources escape the inherent prevention of the working principle of conventional lasers, which naturally relies on the bandgap of semiconductor materials to realize electron-hole recombination. In comparison, the THz photon energy has only several or tens of meV, which is inaccessible to interband transitions. THz-QCLs employ the quantum process of intersubband transitions (ISBT) within repeating superlattice period units [6]. The radiation frequency can be freely tuned solely by engineering the energy separations of dispersed quantum levels, for example, only in the conduction band area. Rapid developments in THz-QCLs have been reported since their first realization in 2002 [5], with the 1.2–5 THz frequency coverage range (operated without the assistance of an external magnetic field [7–9]). Although the initial demonstration of THz-QCLs was performed with the aid of single-plasmon waveguide structures [5], the subsequent development of double-plasmon metal-metal waveguides [10] led to higher operating temperatures (T_{max}). Due to the subwavelength dimensions of the emitting aperture in metal-metal waveguides, their emission is characterized by divergent beam patterns and low output powers [11, 12]. However, unique distributed-feedback (DFB) techniques have recently been demonstrated to achieve single-mode operation with narrow beam patterns [13–15]. Until now, the main limitation for THz-QCL is that the operations suffer from serious thermal degradation, and a room temperature without additional cooling is still not reached. **Figure 1** shows a summary of the improvements in T_{max} over the past decades [16].

The main thermal degradations can be sorted into two types. The first type is the reduction of population inversions (Δn), which can be due to the reduction of population shares in the upper laser state (u, n_u), and an increasing trend at the lower laser state (l, n_l) as the temperature increases. The former can be mainly ascribed to the directly enhanced scattering of the LO-phonon. In general, the active region of the THz-QCL is based on GaAs semiconductor materials, and its LO-phonon energy ($E_{LO} \sim 35$ meV) is quite small. THz radiation is mostly designed below the LO-phonon energy. At high temperatures, the electrons in the upper laser state will acquire enough in-plane kinetic energy (E_{hot}) to relax down to the lower laser state by emitting LO-phonons ($E_{hot} + E_{ul} \approx E_{LO}$) instead of photon emission ($E_p \approx E_{ul}$) [17, 18]. In addition, the up-scatterings from the upper laser state into the higher energy states or directly over the barrier (in the case of low barrier structures) by absorbing the hot phonons [19, 20] lead to efficient carrier leakages. Both of these phonon-assisted processes significantly reduce the population shares in the upper laser state. In addition, the population share of the upper laser state is closely related to the injection efficiency, which can be significantly degraded by injection at high temperatures via

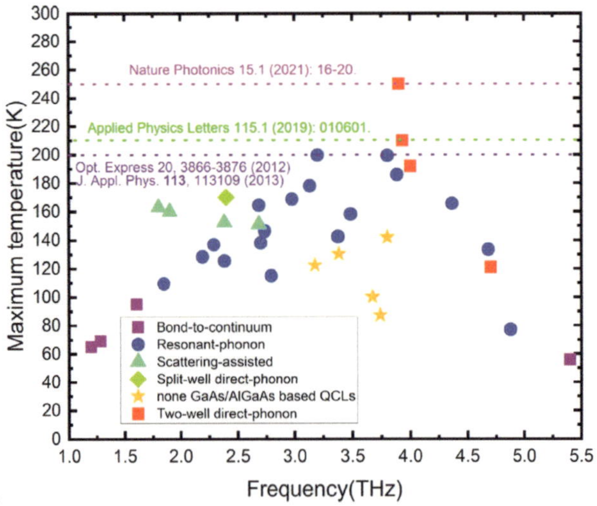

Figure 1.
Survey of the best operating temperature T_{max} at various THz lasing frequency achieved based on different THz-QCL design within pulse operation mode, with the permission from Ref. [16].

resonant tunneling [10, 21]. This is because the resonant tunneling at the injector undergoes enhanced decoherence at high temperatures; as a result, more population is retarded at the injector, thus reducing the population share in the upper laser state. The latter can be thought of as a result of an increased lifetime of the lower laser state because the thermal backfilling from the injector areas is more serious at high temperatures [22, 23]. The second type of degradation is the thermal broadening of the radiation linewidth (Γ) in the active region, and the peak of optical gain is significantly decreased even when a large fraction of population inversion is maintained [24]. This broadening effect is primarily ascribed to the increased Coulomb scattering [25]. Meanwhile, Ref. [25, 26] also pointed out the importance of considering the correlation effect on those broadening for a self-consistent gain profile calculation, which can partially recover the gain peak. These degradation processes occur simultaneously, limiting the final optical gain ($g \sim \Delta n / \Gamma$) below the cavity threshold.

2. THz-QCL designs

The realization of high-T_{max} THz-QCLs is indeed beneficial from the corresponding strategies for suppressing the thermal degradation in the active region and, simultaneously, trying to stabilize the real growth to consistently realize the designs. The multiple-quantum-well active region is usually grown using molecular-beam epitaxy in a $GaAs/Al_xGa_{1-x}As$ material system and is the heart of any QC laser. To obtain the gain for electromagnetic waves at frequency ν, energy levels, wavefunctions, and scattering rates must be properly engineered to provide a population inversion between two states separated by energy $h\nu$. The basic innovation for the successful realization of a QCL can be ascribed to the introduction of the injector region. The optical gain from intersubband transitions in semiconductor superlattices was first investigated early in 1971 [27]; however, early laser proposals suffered from electrical instability owing to the formation of high-field domains crossing the stacked superlattices, until the injector region was particularly defined for offering an accessible operating bias condition, thus realizing the first QCL in the mid-infrared spectrum in 1994, as shown in **Figure 2** [28]. Next, the first THz-QCL lasing at 4.4 THz was experimentally demonstrated in 2002, where most innovations can be ascribed to semi-insulating surface plasmon (SISP) waveguides in the THz spectra [5]. Consequently, the injector has remained a common region in QCLs design for THz QCLs. It should be noted

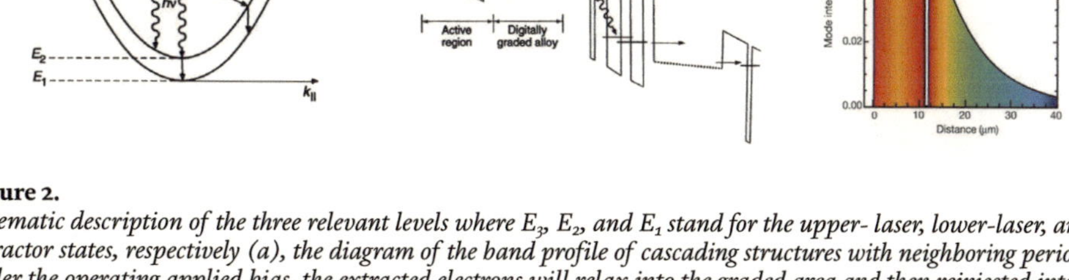

Figure 2.
Schematic description of the three relevant levels where E_3, E_2, and E_1 stand for the upper- laser, lower-laser, and extractor states, respectively (a), the diagram of the band profile of cascading structures with neighboring periods under the operating applied bias, the extracted electrons will relax into the graded area and then reinjected into the upper-laser state in the next period by resonant tunneling (b), Ref. [28]; The semi-insulating surface plasmon (SISP) waveguide for realizing the first THz-QCL (c), where the bottom contact layer is a thin heavy doped GaAs (n^+-GaAs) layer grown on an undoped GaAs substrate, Ref. [5].

that the performance of THz-QCLs is particularly sensitive to the injector because the small energy separation between the laser states makes it difficult to selectively inject carriers (electrons) into the upper laser state [29, 30]. Therefore, improving injection selectivity has gained much attention for different designs [31–33]. The detailed design parameters are trapped in trade-offs from one to another; a successful scheme always requires a dedicated balance of the different parameters.

The main design schemes in the past are shown in **Figure 3a** *chirped superlattice (CSL)*, as shown in **Figure 3a**. The CSL scheme is a main early design, and it employs several quite narrowing quantum wells in which the ground states are raised. When the operating bias is reached, the ground states form "minibands," the lowest state of the pre-minibands is appointed as the upper laser state, and the topmost state in the down-minibands is set as the lower laser state. The electrons are inverted between them and perform the THz photon radiation. Owing to the tight couplings inside the miniband areas (by intra-miniband scattering), electrons always relax quickly to the bottom of the minibands, where the upper laser state is located, the upper laser state is filled, and the lower laser state is depopulated. b) *Bound-to-continuum (BtC)*: As shown in **Figure 3b**, this scheme also employs minibands, and the lower laser state is depleted by intra-miniband scattering, similar to the CSL scheme. The main difference

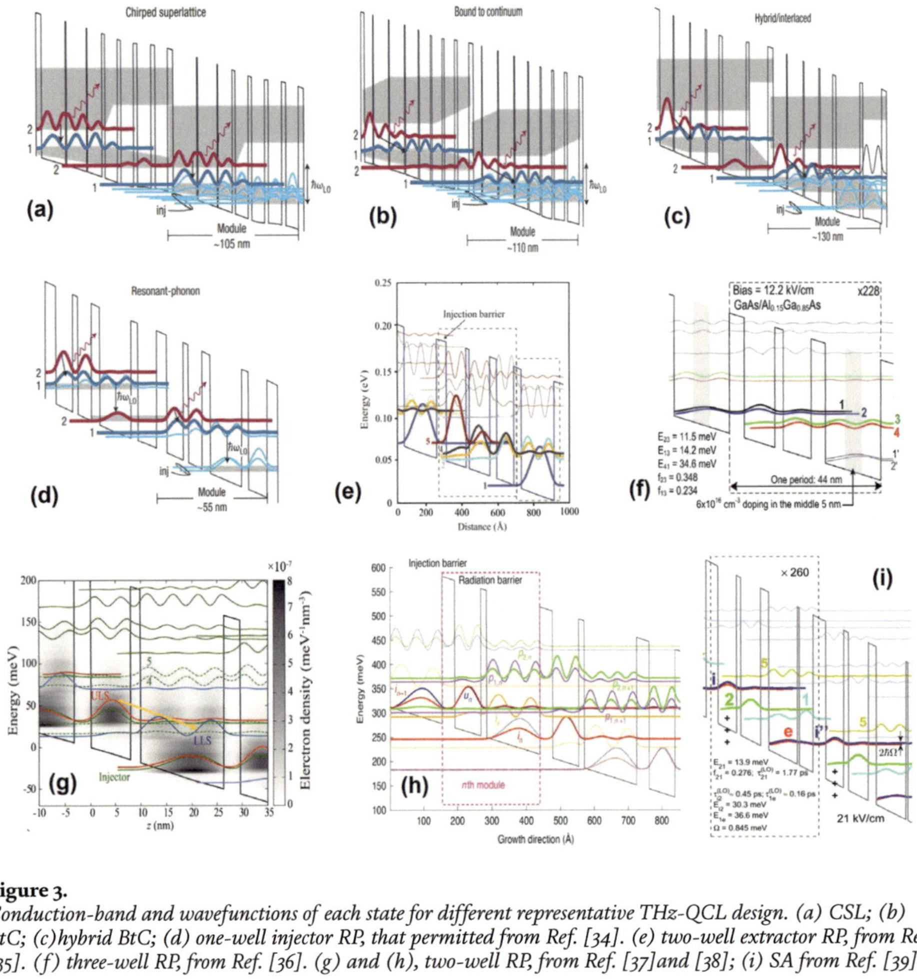

Figure 3.
Conduction-band and wavefunctions of each state for different representative THz-QCL design. (a) CSL; (b) BtC; (c)hybrid BtC; (d) one-well injector RP, that permitted from Ref. [34]. (e) two-well extractor RP, from Ref. [35]. (f) three-well RP, from Ref. [36]. (g) and (h), two-well RP, from Ref. [37]and [38]; (i) SA from Ref. [39].

is in the upper-lower state, where it is positioned diagonally with a lower laser state. This diagonality of radiative transition contributes to an increase in the lifetime of the upper laser state and weakens the parasitic overlap of the wavefunctions between the injector and the lower laser state. Therefore, the BtC design displays an improved T_{max} and output power compared with the CSL design. In addition, an improved BtC scheme with hybrid structures is proposed in which the phonon-assisted depopulation has been incorporated into the miniband areas, which can further relax the thermal backfillings and further improve T_{max} (**Figure 3c**). c) *Resonant-phonon (RP)*: This scheme is specifically indicated for designs in which the injector follows the resonant tunneling mechanism based on a double-level alignment under operating bias. The depopulation process employs LO-phonon resonance, which mostly takes place intra-well. Therefore, the electrons in the lower laser state are scattered down to the injector states by emitting LO-phonons in the sub-picosecond range. There are several RP schemes with different purposes; for example, the one-well injector RP scheme for better injection selectivity by reducing the direct parasitic overlapping between the injector and the lower laser state (**Figure 3d**), or a two-well extractor RP scheme that is similar to the hybrid BtC for offering more tunneling freedom on the design parameters (**Figure 3e**). The first milestone on T_{max} is based on a three-well (4-state) RP scheme (**Figure 3f**), including an adjoining-well resonant-tunneling injection and a one-well extractor where its 1st excited state forms resonance alignment with the lower laser state first, and significantly delocalizes the wavefunction of this lower laser state into the extractor well. Consequently, the lower laser state can form strong coupling with the next injector state with one LO phonon energy downward. The further renewed T_{max} is also based on RP, but with further simplified quantum structures from three-well to two-well (4-state to 3-state, as shown in **Figure 3g** and **h**), where the depopulation from the lower laser state follows only one step via direct intra-well LO phonon scattering. d) *Scattering-assisting (SA)*: As shown in **Figure 3i**, scattering-assisting design is another important scheme. The main difference between SA scheme and the RP design is the injector region, where the SA employs LO-phonon scattering instead of alignment resonance. Resonant tunneling injection always leads to most electrons waiting at the injector before being injected into the upper laser state. Ideally, in the coherent transport regime of resonant tunneling, the upper laser state holds the same number of electrons as the injector state, which is half of the total available population in each period (i.e., a maximum of 50% share in the upper laser state). Owing to the thick injector barrier and the presence of multiple scattering channels, the population share in the upper laser state generally falls below 50%, even at low temperatures. In the SA scheme, by designing the injector to lie one LO-phonon energy up to the upper laser state, this injection limitation can be completely removed.

3. Active-region design by narrowing the period length

Recalling past designs with improved T_{max}, there is a general trend toward reducing the number of states per period (in other words, narrowing the period length by employing a simple quantum structure), and the radiation transition in the active region is strongly diagonal. As shown in **Figure 4**, we followed the structures calculated in Ref. [37] but recalculated them based on our nonequilibrium Green's function (NEGF) model, that is, two-well RP [37, 38], three-well RP [36], four-well SA [39], four-well BtC [35], and seven-well BtC [40]. This indicates an increasing population share in the upper laser state as the total number of states per period is reduced, and the optical gain (the peak) at high temperatures is also significantly improved. Therefore, it provides evidence of the potential of a simplified state system based on the RP scheme. The two-well RP is further optimized by relaxing the shortcomings in its antecessor design of the three-well RP, a) *the two-well RP only uses the resonant tunneling process once at the injector*. Even though three-well RP has been well studied and gives the best T_{max}~ 200 K, a major shortcoming in this scheme is the strict requirement for double resonance alignments at both the injector and extractor region when the bias reaches up to the operating condition. Such a condition is not easy to satisfy in real growth for two reasons. The first reason is the doping issue. High-T_{max} designs generally employ strong diagonality in the active region, where the suitable oscillator strength for radiation transition is only 0.25–0.35. Because the optical gain is linearly related to this strength ($g \propto f$), this small strength should be compensated for by increasing the average doping levels [41]. In THz-QCLs, the dopant position is always selected far away from the radiation area and is generally at the center of the lower well with a doping length of only several nanometers for restraining the dopant migration into the superlattice interfaces. However, this local heavy doping can give rise to severe band bending as a result of space charges, thereby breaking the resonance alignment conditions. Second, active-region growth is based on molecular beam epitaxy (MBE). Although precise atomic controls have been achieved, it is still quite challenging to obtain thousands of thin quantum wells/barriers where the thickness of each layer is only several nanometers, and the percentage variation in the growth rate over the whole growth time (>10 h) can further worsen the double alignment condition [42]. In addition, the optical gain profile can be broadened and deformed because the applied bias can closely affect the couplings in the active region; thus, the exact laser frequencies suffer from uncertainties [21]. b). *The depopulation in a two-well RP follows an intra-well direct-phonon scattering*. The intra-well

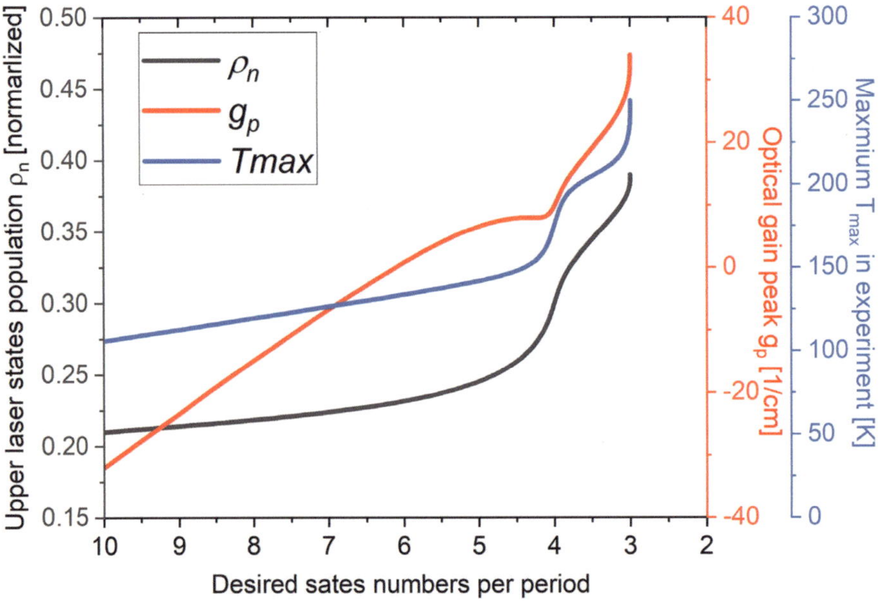

Figure 4.
Calculated population share the upper-laser state (ρ_n), the calculated optical gain peak g_p, and the T_{max} in experiments. The calculations are based on the designs of two-well (3-state) RP [37, 38], three-well (4-state) RP [36], four-well(4-state) SA [39], and four-well(5-state) BtC [35], seven-well(10-state) BtC [40].

resonance phonon scattering is performed between its 1st excited and ground states in the lower-well, and there is a very strong coupling between them that guarantees a highly efficient depopulation within ultrafast times. Although the energy separation between them can affect depopulation efficiency, it maintains high robustness over a large tuning range of energy separation (30~60 meV [43]). c). *The barriers in two-well RP are much taller*. The tall barrier, by using $Al_{0.25}Ga_{0.75}As$ [37] and $Al_{0.3}Ga_{0.7}As$ [38], can confine the electrodes deeply at the desired states; thus, suppressing the carriers leaking from the confined states over the barriers into the continuum at high temperatures. In a tall barrier design case, careful quantum structure engineering for clean n-clean systems can further minimize the thermal up-scattering from the desired states into the high-lying nonrelevant states. d). *The two-well RP intentionally uses larger depopulation energy for the LO-phonon resonance*. The depopulation energy at the extractor in the high-T_{max} two-well RP is 42 meV in Ref. [37] and 52 meV in Ref [38], which is higher than the previous designs with that exactly equal to one LO-phonon of 36 meV. Although Ref. [38] emphasized the benefits of nonrelevant high-confined state engineering, the contribution of the higher depopulation energy to suppress thermal backfilling is at least equal. In addition to the RP scheme, the narrow-period concept is also attractive for the SA scheme; however, the SA scheme based on a two-well structure has still not been realized in experiments.

4. Limitations in narrow-period active-region design (RP and SA scheme)

4.1 Two-well RP design

4.1.1 Intra-period thermal up-scattering via non-relevant states

The significance of the high-lying nonrelevant states on high-T_{max} has been demonstrated [19, 20, 44] and indicates that the up-scatterings that originate from the upper laser state into those nonrelevant states in the same period will play a significant role at high temperatures. The up-scattering activation energy E_a was extracted by fitting the Arrhenius plots from the measured device output power by [44]:

$$\ln\left(1-\frac{P_{out}(T)}{P_{out(\max)}}\right)\approx\ln(a)-\frac{E_a}{KT} \tag{1}$$

As shown in **Figure 5**, an E_a of 35 meV is extracted, which presents the thermal up-scattering occurring between the upper laser state and the 1st nonrelevant in the lower well (the 2nd excited state from this well). Following this up-scattering channel, owing to the relatively low activation energy, as the temperature increases, the electrons can escape into the continuum quickly because this nonrelevant state is close to the continuum, which results in a sharp increase in J_{th} at temperatures above 100 K (**Figure 5f**). Therefore, a tall barrier is essential to confine the up-scattered electrons deeply or to increase the up-scattering activation energy. The validity of a tall barrier for suppressing such channels is shown in **Figure 5f** and **g** ($As_{0.15}Ga_{0.85}As \rightarrow Al_{0.3}Ga_{0.7}As$), where a much higher up-scattering activation energy is extracted, that is, 35→117 meV. The corresponding J_{th} is much lower at high temperatures, and T_{max} improves from 121 to 173 K.

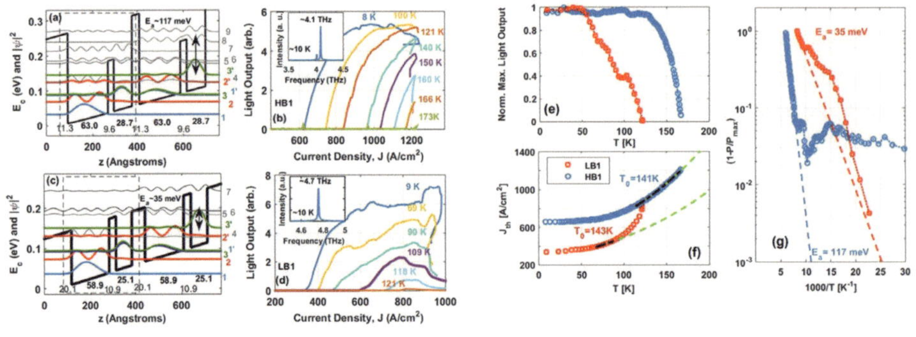

Figure 5.
Band diagram of the two-well RP design by tuning the barrier height. The corresponding light output at different temperatures are shown (a–d). The Arrhenius plots with the fitting activation energy and the J_{th} depending on the temperature is also shown (e–g), Ref. [44].

4.1.2 Depopulation energy and thermal backfilling

Ref. [45] shows that at high temperatures, the thermal backfilling from the injector into the lower-laser state is the main source of the population in the lower-laser state; therefore, dramatically reducing the population inversion. The thermal backfilled electrons (n_{bf}) approach 20% of the population share at 300 K, as shown in **Figure 6a** and **b**, which almost equals to the population of the upper-laser state population; thus, non-inversion can be formed. There are two reasons for this: First, the RP scheme relies on resonant tunneling for electrons injected from the injector, where most of the population resides at the injector; therefore, playing the role of an electron pool for thermal backfilling. Second, the GaAs material has a small LO-phonon of 36 meV. The RP design normally uses depopulation energy that is equal to one LO-phonon; this inherently small depopulation energy encourages the occurrence of thermal backfilling. This also indicates the significance of exploring large LO-phonon energy materials such as GaN [46] and ZnO [47] to suppress thermal backfilling.

If the designs are still based on the GaAs material, enlarging the depopulation energy is an effective method for suppressing the thermal backfilling, and large depopulation energy can further reduce the parasitic coupling of the upper-laser state

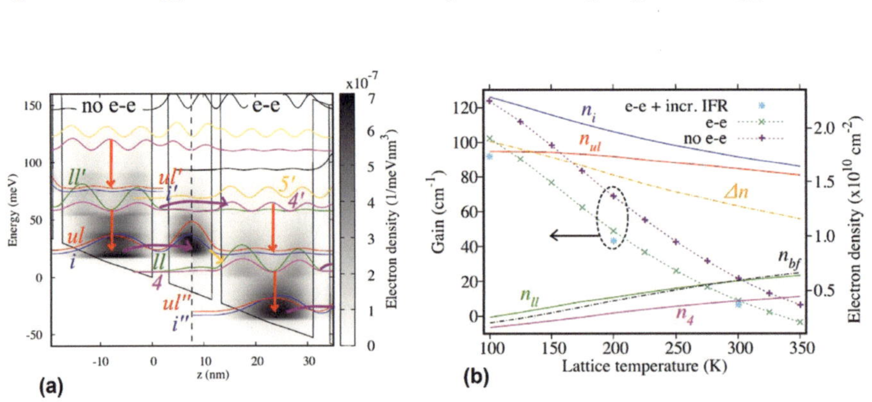

Figure 6.
Band diagram in two-well RP design where the population distribution at each state is given at high temperatures(a); The thermal backfilled population from the injector into the lower-laser state (n_{bf}) is particularly (b), Ref. [45].

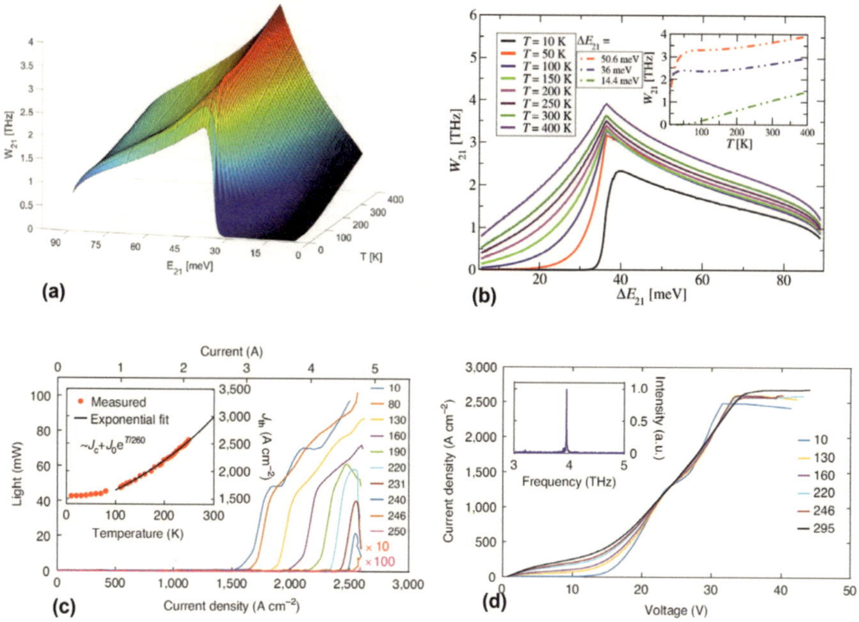

Figure 7.
Averaged electron-LO phonon scattering rate between the 1st excited state and the ground state by varying the energy separation between them (a); A representative cross-section of the figure a (b), Ref. [43]; Plot of light output versus current density at different temperatures. Inset: plot of threshold current density versus temperature (J_{th}–T) with J_c and J_o as fitting parameters. The black line is an exponential fit (c); Plot of voltage versus current density (J–V) at different temperatures. Inset: a lasing spectrum taken at 246 K (d), Ref. [38].

directly with the next injector; therefore, protecting the lifetime of the upper-laser state by preventing parasitic LO-phonon scattering between them. However, the depopulation energy deviating from one LO-phonon slows down the depopulation process, thus increasing the lifetime of the lower-laser state and reducing the population inversion. Ref. [43] estimated the average scattering rate for intra-well LO phonon resonance between the 1st and 2nd states in a single well based on a self-consistent Schrödinger–Poisson solver (**Figure 7a** and **b**). The exact energy separation between them was tuned by varying the well width. It is evident that at a high temperature of 300 K, the depopulation efficiency (here simply assumed by the scattering rate) is reduced by 33% when large depopulation energy of 60 meV is selected. Considering the highest T_{max} in Ref. [38] (**Figure 7c** and **d**), they employed a larger depopulation energy of 52 meV and the improved characteristic temperature of J_{th} in this Ref. was also largely ascribed to the suppression of thermal backfilling, together with the thermal up-scatterings via high-lying nonrelevant states.

4.1.3 Interperiod interactions via nonrelevant states

The two-well configuration inherently has a narrow period length such that the parasitic interactions in neighboring periods can be more serious than in the wide-period design. Furthermore, a narrow period also leads to a stronger electrical field that will lower the nonrelevant states in energy downstream. Ref. [48] addresses the critical effect of nonrelevant high-lying nonrelevant states on the laser threshold current by forming a resonant-tunneling-like channel between three neighboring periods. The

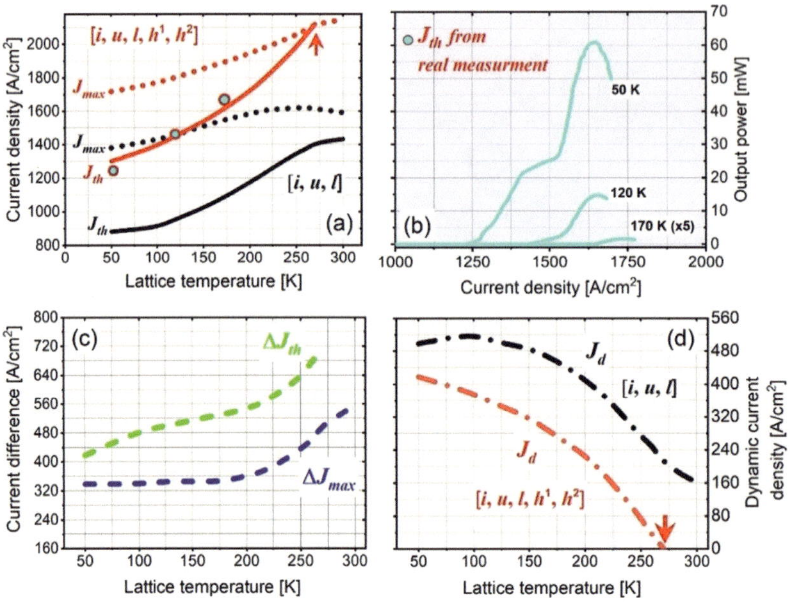

Figure 8.

J_{max} and J_{th} depend on the temperatures for only desired states [i, u, l] (black curves) and more high-lying nonrelevant states [i, u, l, h^1, h^2] (red curves) (a); The light output measured devices based on this 2-well RP design. The J_{th} at different temperatures are extracted from those lasing curves and are shown in figure a (b); The changing magnitude of J_{max} and J_{th} (ΔJ_{max}, ΔJ_{th}) by including or excluding the nonrelevant states in the calculations(c); The laser dynamic of J_d depending on the temperatures in both the states inclusion(d), Ref. [48].

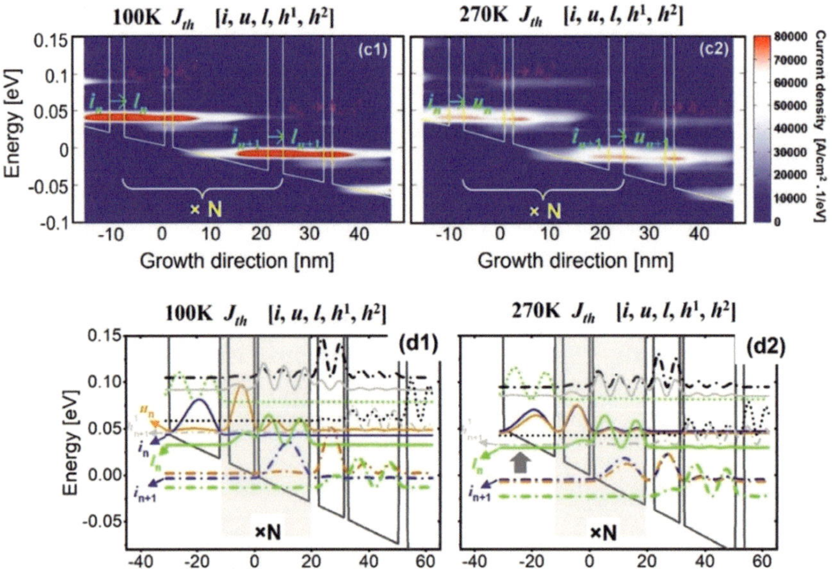

Figure 9.

Threshold current J_{th} mappings that resolved by growth direction and energy at both low and high temperatures (100 K, 270 K) for the desired states [i, u, l] and more high-lying nonrelevant states included [i, u, l, h^1, h^2], Ref. [48].

laser dynamics in current will significantly shrink to zero even at 270 K (**Figure 8**). The spatial and energy-resolved current mapping between the neighboring periods clearly shows the tunneling current along the growth directions (**Figure 9**).

4.2 Two-well SA design

4.2.1 More serious inter-period interactions via irrelevant states

The two-well configuration for the SA scheme shows the main difference is that its upper-well is wider for an intra-well LO-phonon scattering injection, and its lower-well is also wide enough to ensure the lower-laser state down to the upper laser state for THz radiation. Therefore, the high-lying nonrelevant states in the lower well were inherently low in energy, as shown in **Figure 10a**. The current-voltage plots in **Figure 10b** show that, at the operating bias, the peak current increases appreciably when both nonrelevant states 4 and 5 are allowed in the calculation. The inclusion of more high-energy states (nonrelevant states 6, 7, and 8) increases the current density further, but only slightly. Therefore, nonrelevant states 4 and 5, which belong to different neighboring periods, are crucial for serious current leakage. It is visible in the spatial and energy resolved current density mappings in **Figure 10c** (A, B, C, D), and the leakage channel is formed due to a strong interaction of the upper-laser states in the upstream period (state $2n_{-1}$) with state 4_n in period n and sequential tunneling to state $5n_{+1}$ in the downstream period n+1 because three of these states are close both in energy and space.

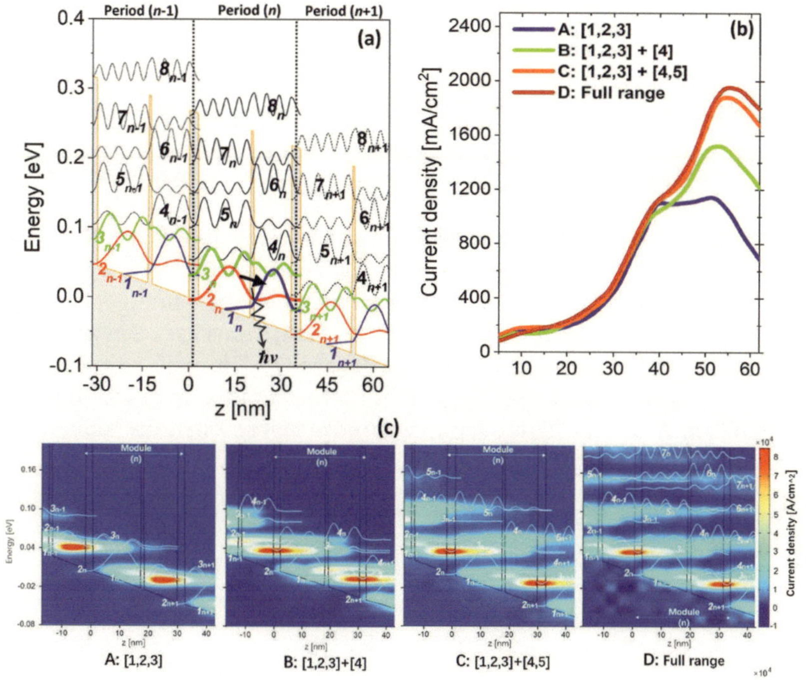

Figure 10.
Conduction band diagram and the tight-binding states of two-well SA design with three neighboring periods (n-1,n,n+1) (a); The voltage-current plots at high temperatures of 300 K when the different number of the high-lying nonrelevant states are included for calculations (b); The current mapping resolved by the growth direction and the energy when the nonrelevant states are included (c), Ref. [49].

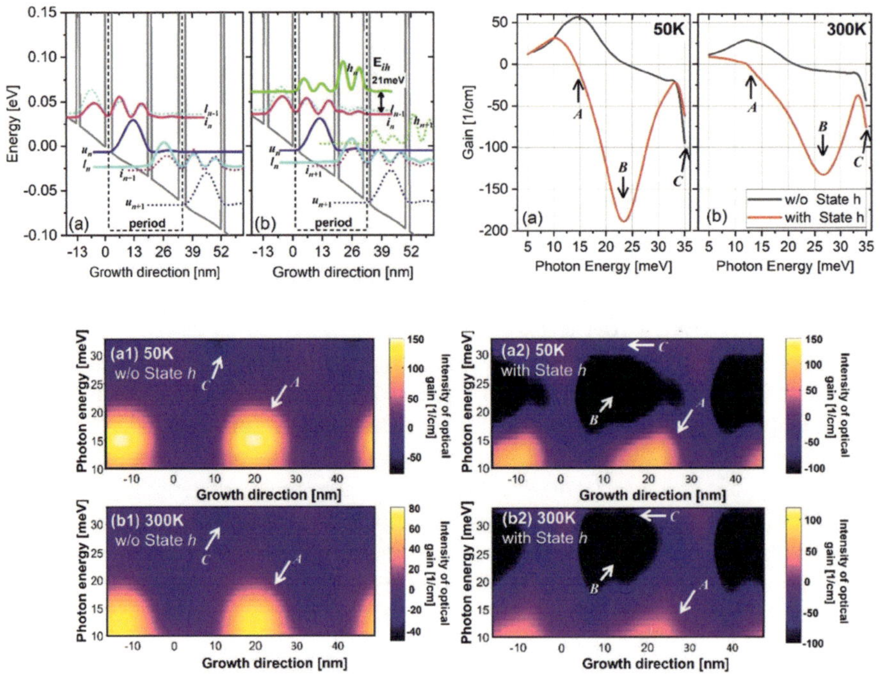

Figure 11.
Band diagram of the two-well SA design in cases of the desired states and the critical nonrelevant state. The optical gain spectra under two different number of stated inclusions. The spatial and energy resolved gain clearly shows the overlapping of the parasitic absorption and the gain area, Ref. [50].

4.2.2 Parasitic absorption overlapping gain

In the two-well SA design (**Figure 11**), there is a strong decoupling between population inversion and optical gain, in which the population inversion does not change significantly when the high-lying nonrelevant states are included, but the peak of gain is significantly limited by those high-lying states. The reduction of the gain peak is obvious at a low temperature of 50 K. This was ascribed to the emergence of inter-period parasitic absorption, which was caused by transitions between the injector and the first nonrelevant states in the lower well (**Figure 11**). This parasitic absorption unavoidably overlaps with the optical gain due to the engineering limit permitted in a simple quantum structure. This overlap was more severe when the lasing frequency exceeded 3 THz in the two-well SA design.

5. Further narrow-period designs

5.1 Split-well three-state RP design

The split-well direct-phonon concept described in Ref. [51] was proposed by using only the ground states to push up the nonrelevant states for a clean-state system, while keeping the depopulation energy almost equal to 36 meV for ultrafast extraction. In this design, each period contained three wells and employed the ground states in each well. Due to all three wells being narrow, the total periodic length is still quite

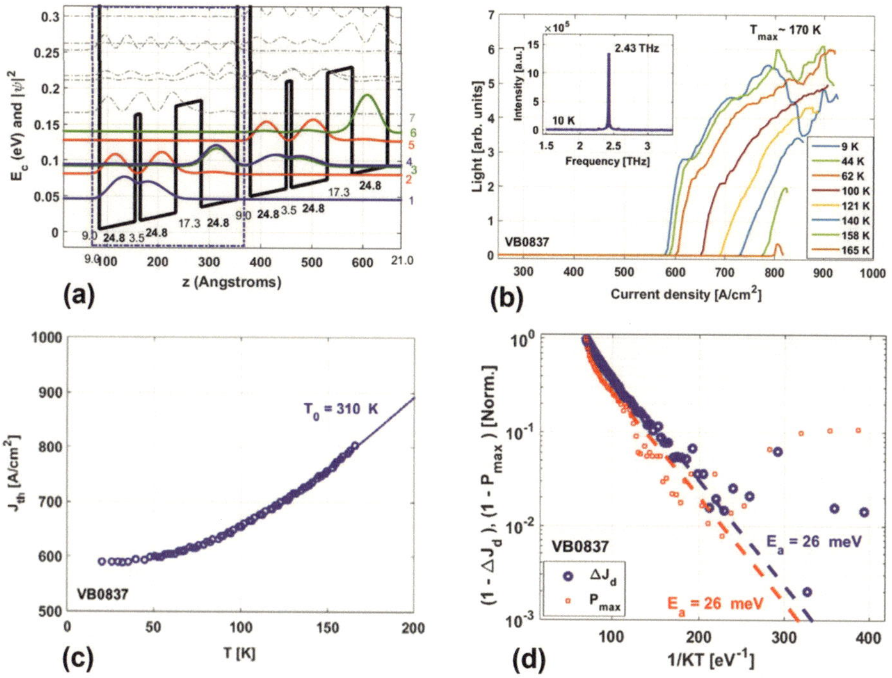

Figure 12.
Band profile and the wavefunctions in the "split-well" design (a); Pulsed light output versus current measurement(b); The threshold current vs temperature(c); Activation energy extracted from the laser's maximum output power (P_{max}) vs temperature, and the current dynamic range $\Delta J_d = (J_{max} - J_{th})$ vs temperature (d), Ref. [51].

small as a "narrow-period" design. Compared with the "all-ground" design from Ref. [52], the main advance in this split-well design is the very strong depopulation couplings between the lower-laser state and the injector, where the "all-ground" design follows a diagonal depopulation process and suffers from limited laser dynamics. As shown in **Figure 12**, the results demonstrate that "split-well" lasers profit from both eliminations of up-scatterings via high-lying nonrelevant states and resonant depopulation of the lower laser states. A negative differential resistance was observed at room temperature beyond the operating bias conditions, indicating that the state system performs as a clean 3-state system, where a high characteristic temperature of J_{th} is fitted out. However, the difficulty of this design is that the splitting at the extractor closely relies on the alignment, and the ultrathin extractor barrier, which is only 3.5 monolayer, is not easily obtained consistently during the long-time MBE growth.

5.2 Two-well double-phonon RP designs

The design in Ref. [53] mainly considers the effect of thermal backfilling on the lower-laser state by using large depopulation energy of more than one LO phonon. This is different from previous designs; the lower-laser state is appointed by the 2nd excited state in the lower well. Simultaneously, the 1st excited state from the lower well is also a depopulation destination, and as a result, three transitions are responsible for the depopulation, that is, 3→2, 3→1, and 2→1, as shown in **Figure 13a**. The tuning of the energy separation between 3 and 2, E_{32}, and also between 2 and 1, E_{21}, was studied, revealing that only by maintaining the total energy separation

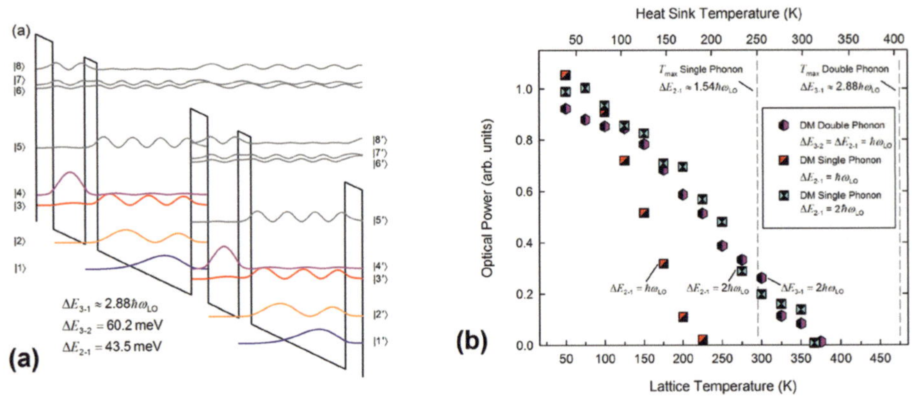

Figure 13.
Conduction band diagram and the wavefunctions in a double phonon intrawell depopulated based on a two-well structure(a); The calculated optical output versus temperature by tuning the depopulation energy with single- and double-phonon(b), Ref. [53].

between the lower laser state and the next injector, E_{31}, more than a double phonon, the thermal backfilling can be well suppressed and thus shows a cut-off temperature above 300 K (**Figure 13b**). The challenge with such large depopulation energy is that it requires a much taller barrier, and the operating electric field is very strong, which causes many difficulties in device realization.

5.3 Two step-well SA design

As described above, in a two-well SA design, the high-lying nonrelevant state (the 1st excited state in the lower well), h, as shown in **Figure 14a**, is naturally close to the desired states and forms a serious parasitic absorption overlapping the gain

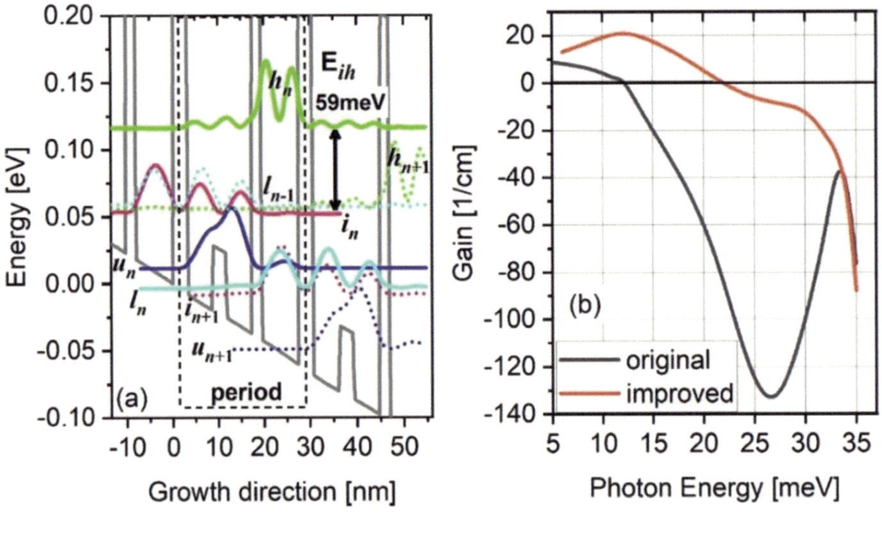

Figure 14.
Improved two-well SA design by employing ministeps ($Al_{0.05}Ga_{0.95}As$) where the tall barriers of $Al_{0.05}Ga_{0.95}As$ are used(a); Comparison of optical gain spectra at 300 K in this improved design (red curve) and the regular two-well SA design. Both spectra are predicated by considering the high-lying nonrelevant states (u, l, i, h) (b), Ref. [50].

areas. Consequently, the peak gain was largely limited even at low temperatures. Ref. [50] attempted to narrow the upper well, where the lower well can be correspondingly narrower. Therefore, the irrelevant state *h* moves upward. To satisfy the energy separation between injector states *i* and the upper laser state *u* close to the LO-phonon, a ministep was intentionally introduced into the upper well. By simply tuning this ministep, the energy separation of the injector state *i* and the upper-laser state *u* can be maintained at 36 meV. To achieve the same THz frequency of 3.7 THz, the upper well is narrowed by 13% compared with the regular two-well SA design (16.2→14.2 nm), and the lower well is also narrowed by 33% (10.6→8.18 nm). The irrelevant state *h* is thus effectively moved upward, and the energy separation E_{ih} is significantly increased from 21 to 59 meV. The optical gain spectra in **Figure 14b** show complete suppression of overlapping from this parasitic absorption and recovery of the peak gain.

6. Strict request on MBE growth controls for the narrow-period designs

The simple structure requires more precise control of the growth because each layer performs multiple functions as the design parameters. The recent highest T_{max} emphasizes high-quality growth on repeating periods and interfaces. It is very challenging to grow thin layers, especially in the case of tall barriers with several monolayers. The interface flatness, doping controls, and the period repeated during the long-time growth are tough topics for start-of-art designs. Meanwhile, uniformity across a 3-inch substrate is also critical, even for optimizing the rotations carefully, and an arbitrary thickness error can still be approximately 2%. Reliable and repeatable growth is essential for dedicated design parameters for a higher T_{max} (i.e., even within a very small range of parameter balances in such designs).

7. Summary

Despite the extensive application potential of the THz spectrum, light radiation sources remain the most urgent, where the most expected is a compact solid-state device analogous to conventional semiconductor lasers. The THz quantum cascade laser has been treated as the most attractive candidate to fill the THz gap and was soon realized in the THz range after the experimental demonstration in the mid-infrared range. However, this type of THz laser suffers from thermal degradation that cannot work at room temperature until now. Different models attempt to describe the quantum transport in THz-QCLs and clarify the thermal limitations, then predict the designs with high-temperature tolerances. The design for high-temperature operation follows a path by simplifying the containing quantum well structure in each period; however, such a narrow period requires careful engineering of the barrier and requires the control of the high-lying nonrelevant states to suppress any inter-period parasitic channels (for example, creating a clean state system). Different strategies are needed depending on the injection method (resonant tunneling method or scattering-assisted method). In addition, to realize these narrow-period structures, precise control of the MBE growth is essential to ensure the accurate thickness of each layer and the flatness of interfaces with uniform alloyed barriers.

Author details

Li Wang* and Hideki Hirayama
Research Center for Advanced Photonics, RIKEN, Sendai, Japan

*Address all correspondence to: li.wang@riken.jp

References

[1] Mittleman DM. Twenty years of terahertz imaging. Optics Express. 2018;**26**:9417-9431

[2] Siegel PH. Terahertz technology. IEEE Transactions on Microwave Theory and Techniques. 2002;**50**:910-928

[3] Korter T, Plusquellic DF. Continuous-wave terahertz spectroscopy of biotin: Vibrational anharmonicity in the far-infrared. Chemical Physics Letters. 2004;**385**:45-51

[4] Ogawa Y, Hayashi S, Oikawa M, Otani C, Kawase K. Interference terahertz label-free imaging for protein detection on a membrane. Optics Express. 2008;**16**:22083-22089

[5] Kohler R, Tredicucci A, Beltram F, Beere HE, Linfield EH, Davies AG, et al. Terahertz semiconductor-heterostructure laser. Nature. 2002;**417**:156-159

[6] Williams BS, Callebaut H, Kumar S, Hu Q, Reno JL. 3.4-THz quantum cascade laser based on longitudinal-optical-phonon scattering for depopulation. Applied Physics Letters. 2003;**82**:1015-1017

[7] Luo H, Laframboise SR, Wasilewski ZR, Aers GC, Liu HC, Cao JC. Terahertz quantum cascade lasers based on a three-well active module. Applied Physics Letters. 2007;**90**:041112

[8] Kumar S, Lee AWM. Resonant-phonon terahertz quantum-cascade lasers and video-rate terahertz imaging. IEEE Journal of Selected Topics in Quantum Electronics. 2008;**14**:333-344

[9] Scalari G, Walther C, Fischer M, Terazzi R, Beere H, Ritchie D, et al. THz and sub-THz quantum cascade lasers. Laser & Photonics Reviews. 2009;**3**:45-105

[10] Callebaut H, Hu Q. Importance of coherence for electron transport in terahertz quantum cascade lasers. Journal of Applied Physics. 2005;**98**:104505

[11] Kohen S, Williams BS, Hu Q. Electromagnetic modeling of terahertz quantum cascade laser waveguides and resonators. Journal of Applied Physics. 2005;**97**:053106

[12] Adam AJL, Kasalynas I, Hovenier JN, Klaassen TO, Gao JR, Orlova EE, et al. Beam patterns of terahertz quantum cascade lasers with subwavelength cavity dimensions. Applied Physics Letters. 2006;**88**:151105

[13] Fan JA, Belkin MA, Capasso F, Khanna S, Lachab M, Davies AG, et al. Surface emitting terahertz quantum cascade laser with a double-metal waveguide. Optics Express. 2007;**14**:11672-11680

[14] Kumar S, Williams BS, Qin Q, Lee AWM, Hu Q, Reno JL. Surface-emitting distributed feedback terahertz quantum-cascade lasers in metal-metal waveguides. Optics Express. 2007;**15**:113-128

[15] Chassagneux Y, Colombelli R, Maineult W, Barbieri S, Beere HE, Ritchie DA, et al. Electrically pumped photonic-crystal terahertz lasers controlled by boundary conditions. Nature. 2009;**457**:174-178

[16] Wen BY, Ban DY. High-temperature terahertz quantum cascade lasers. Progress in Quantum Electronics. 2021;**80**:100363

[17] Albo A, Hu Q. Investigating temperature degradation in THz quantum cascade lasers by examination

of temperature dependence of output power. Applied Physics Letters. 2015;**106**:131108

[18] Kumar S, Hu Q, Reno JL. 186 K operation of terahertz quantum-cascade lasers based on a diagonal design. Applied Physics Letters. 2009;**94**:131105

[19] Albo A, Hu Q. Carrier leakage into the continuum in diagonal GaAs/$Al_{0.15}$GaAs terahertz quantum cascade lasers. Applied Physics Letters. 2015;**107**:241101

[20] Albo A, Hu Q, Reno JL. Room temperature negative differential resistance in terahertz quantum cascade laser structures. Applied Physics Letters. 2016;**109**:081102

[21] Kumar S, Hu Q. Coherence of resonant-tunneling transport in terahertz quantum-cascade lasers. Physical Review B. 2009;**80**:245316

[22] Williams BS, Kumar S, Qin Q, Hu Q. Terahertz quantum cascade lasers with double-resonant-phonon depopulation. Applied Physics Letters. 2006;**88**:261101

[23] Lü JT, Cao JC. Coulomb scattering in the Monte Carlo simulation of terahertz quantum-cascade lasers. Applied Physics Letters. 2006;**88**:061119

[24] Nelander R, Wacke A. Temperature dependence of the gain profile for terahertz quantum cascade lasers. Applied Physics Letters. 2008;**92**:081102

[25] Grange T. Contrasting influence of charged impurities on transport and gain in terahertz quantum cascade lasers. Physical Review B. 2015;**92**:241306

[26] Banit F, Lee S-C, Knorr A. Self-consistent theory of the gain linewidth for quantum-cascade lasers. Applied Physics Letters. 2005;**86**:041108

[27] Kazarinov RF, Suris RA. Possibility of the amplification of electromagnetic waves in a semiconductor with a superlattice. Soviet Physics – Semiconductors. 1971;**5**:707-709

[28] Faist J, Capasso F, Sivco DL, Sirtori C, Hutchinson AL, Cho AY. Quantum cascade laser. Science. 1994;**264**:553-556

[29] Kumar S, Williams BS, Hu Q, Reno JL. 1.9 THz quantum-cascade lasers with one-well injector. Applied Physics Letters. 2006;**88**:121123

[30] Walther C, Scalari G, Faist J, Beere H, Ritchie D. Low frequency terahertz quantum cascade laser operating from 1.6 to 1.8-THz. Applied Physics Letters. 2006;**89**:231121

[31] Walther C, Fischer M, Scalari G, Terazzi R, Hoyler N, Faist J. Quantum cascade lasers operating from 1.2 to 1.6 THz. Applied Physics Letters. 2007;**91**:131122

[32] Williams BS, Kumar S, Hu Q. Operation of terahertz quantum-cascade lasers at 164 K in pulsed mode and at 117 K in continuous-wave mode. Optics Express. 2005;**13**:3331-3339

[33] Khanal S, Reno JL, Kumar S. 2.1 THz quantum-cascade laser operating up to 144 K based on a scattering-assisted injection design. Optics Express. 2015;**23**:19689-19697

[34] Williams BS. Terahertz quantum-cascade lasers. Nature Photonics. 2007;**1**:517-525

[35] Ohtani K, Turčinková D, Bonzon C, Beck M, Faist J, Justen M, et al. High performance 4.7 THz GaAs quantum cascade lasers based on four quantum wells. New Journal of Physics. 2016;**18**:12300

[36] Fathololoumi S, Dupont E, Chan CWI, Wasilewski ZR, Laframboise SR, Ban D, et al. Terahertz quantum cascade lasers operating up to~ 200 K with optimized oscillator strength and improved injection tunneling. Optics Express. 2012;**20**:3866-3876

[37] Bosco L, Franckié M, Scalari G, Beck M, Wacker A, Faist J. Thermoelectrically cooled THz quantum Cascade laser operating up to 210 K. Applied Physics Letters. 2019;**115**:010601

[38] Khalatpour A, Paulsen AK, Deimert C, Wasilewski ZR, Hu Q. High-power portable terahertz laser systems. Nature Photonics. 2021;**15**:16-20

[39] Razavipour SG, Dupont E, Fathololoumi S, Chan CWI, Lindskog M, Wasilewski ZR, et al. An indirectly pumped terahertz quantum cascade laser with low injection coupling strength operating above 150 K. Journal of Applied Physics. 2013;**113**:203107

[40] Scalari G, Hoyler N, Giovannini M, Faist J. Terahertz bound-to-continuum quantum-cascade lasers based on optical-phonon scattering extraction. Applied Physics Letters. 2005;**86**:181101

[41] Chan CWI, Albo A, Hu Q, Reno LJ. Tradeoffs between oscillator strength and lifetime in terahertz quantum cascade lasers. Applied Physics Letters. 2016;**109**:201104

[42] Andrews AM, Zedeerbauer T, Hermann M, MacFarland DS, Strasser G. Quantum Cascade Lasers, Molecular Beam Epitaxy from Research to Mass Production. Amsterdan, Nertherlands: ScienceDirect Book; 2018. pp. 597-624

[43] Demic A, Ikonic Z, Dean P, Indjin D. Prospects of temperature performance enhancement through higher resonant phonon transition designs in GaAs-based terahertz quantum-cascade lasers. New Journal of Physics. 2022;**24**:033047

[44] Albo A, Flores YV, Hu Q, Reno JL. Two-well terahertz quantum cascade lasers with suppressed carrier leakage. Applied Physics Letters. 2017;**111**:111107

[45] Frankie M, Bosco L, Beck M, Bonzon C, Mavrona E, Scalari G. Two-well quantum cascade laser optimization by non-equilibrium Green's function modelling. Applied Physics Letters. 2018;**112**:021104

[46] Wang L, Lin T-T, Chen M-X, Wang K, Hirayama H. Engineering of electron–longitudinal optical phonon coupling strength in m-plane GaN terahertz quantum cascade lasers. Applied Physics Express. 2021;**14**:112003

[47] Bellotti E, Driscoll K, Moustakas TD, Paiella R. Monte Carlo simulation of terahertz quantum cascade laser structures based on wide-bandgap semiconductors. Journal of Applied Physics. 2009;**105**:113103

[48] Wang L, Lin T-T, Wang K, Hirayama H. Nonrelevant quantum levels limited laser dynamic in narrow-period terahertz quantum cascade lasers Li Wang. Japanese Journal of Applied Physics. 2022

[49] Wang L, Lin T-T, Wang K, Grange T, Birner S, Hirayama H. Short-period scattering-assisted terahertz quantum cascade lasers operating at high temperatures. Scientific Reports. 2019;**9**:9446

[50] Wang L, Lin T-T, Wang K, Hirayama H. Limitation of parasitic absorption in designs of three-state terahertz quantum cascade lasers with direct-phonon injection. Applied Physics Express. 2022;**15**:052002

[51] Albo A, Flores YV, Hu Q, Reno JL. Split-well direct-phonon terahertz quantum cascade lasers. Applied Physics Letters. 2019;**114**:191102

[52] Chan CWI, Hu Q, Reno JL. Ground state terahertz quantum cascade lasers. Applied Physics Letters. 2012;**101**:151108

[53] Freeman W. Double longitudinal-optical phonon intrawell depopulated terahertz quantum cascade structures: Electron transport modeling using a density matrix method. Applied Physics Letters. 2021;**118**:241107

Chapter 4

Degradation Analysis of Silicone as Encapsulation and Molding Material in High Power LEDs

Abdul Shabir and Cher Ming Tan

Abstract

Applications of LEDs have increased significantly, and increasing outdoor applications are observed. Some outdoor applications require high reliability as their failure can lead to hazardous consequences. Examples are their applications in automotive, street lamp lighting etc. To ensure the reliability of LEDs in outdoor applications, reliability test that include humidity on the LEDs must be done. However, it is found that accelerated life test of LEDs at high humidity level cannot be extrapolated to standard condition of lower humidity as the mechanism of degradation depends critically on humidity level. In fact, the degradation of LEDs in outdoor applications is mainly due to the degradation of their encapsulation and housing materials (or called packaging material as a whole) instead of the semiconductor chip itself. The decrease in lumen is the results of crack and discoloration of the LED packaging material. Detail understanding of the failure physics of the packaging material for LED under humidity is needed for extrapolation performed at accelerated stress condition so that LED luminary reliability can be predicted. This chapter reviews the different types of degradation physics of the packaging material using ab-initio simulation with excellent verification from experiments. The method of extrapolation is therefore derived from the physics-based model after detailed understanding of the degradation physics of LEDs. The model also provides strategy for industry to prolong the usage of LEDs in outdoor applications, either through materials or operating conditions selection.

Keywords: light emitting diodes, failure analysis, LED degradation, silicone, PDMS, density functional theory

1. Introduction

Light is one of the most essential natural elements that has helped humans since early ages, and our dependency on light is very high. With the development of artificial modes of electrically powered lighting and heating devices for indoor as well as outdoor applications, the demand of electrical power now more than ever. About one-fifth of the total power is used for lighting [1]. To reduce power usage while maintaining the standard of complexity in lifestyle as well as advancement of technologies, the search for reliable and cost-effective light sources is in demand more than ever.

Light sources	Illumination efficiency (1 m/W)	Lifetime (hours)
Incandescent	15–20	1000
Tungsten Halogen	12–35	2000–4000
Mercury Vapor	40–60	12,000
Lamps Compact Fluorescent	40–70	6000–12,000
Lamp Fluorescent Lamp	50–100	10,000–16,000
Induction Lamp	60–80	60,000–100,000
High Pressure Sodium	80–100	12,000–16,000
LED	80–160	50,000–100,000

Table 1.
Lighting efficiency and lifetime of some light sources [1, 2].

With the discovery of high power light emitting diode (LED), LEDs have become the go-to sources for applications in lighting systems. Due to their prolonged lifetime, low maintenance, small sizes and energy efficient properties, LEDs have replaced most of the conventional sources of lighting such as incandescent and fluorescent lamps all over the world. The technology used in LEDs is solid-state lighting (SSL) which provides more energy saving than conventional fluorescent lamps. The comparative study between efficiency and power output of LEDs and other conventional sources of lighting is shown in **Table 1** [1, 2].

Apart from energy efficiency, LEDs also have a longer lifetime compared to fluorescent and incandescent lamps. These devices are compact and have higher mechanical resistance compared to conventional sources. The report by Guan et.al showed nanowire LEDs that were able to bent for −2.5 mm to +3.5 mm radii of curvature while the light intensities was consistent throughout the bending cycle tests [3]. If all the conventional light sources in the world were converted to the LEDs, energy consumption could be reduced by 230 typical 500-MW coal plants, and greenhouse gas emission can then be reduced by about 200 million tonnes [4]. Owing to these benefits, high-power LEDs are now not only used to replace incandescent and halogen lamps in houses, they are also used in outdoor lighting applications such as displays, automobiles, traffic signals, lamp posts, emergency medical services, marine applications, etc. [4–12]. Unfortunately, degradation in LEDs will not only exhibit reduced light output but also change in color which can affect some of their applications. For example, a dimmed LEDs in traffic light might cause road accident to happen. Sudden blackout due to the failure of LEDs in lamp posts can create danger to the pedestrians, and these have indeed resulted in lawsuits against the manufacturers [13–16]. Thus, alongside with energy efficiency and conversion, the reliability of LEDs is of the utmost importance [17], especially our dependence on the lighting is so high.

The environmental stresses for LEDs in outdoor applications, such as humidity and temperature can lead to degradation of output light from LEDs. While reliability tests of LEDs can help to evaluate if they are suitable for various applications, a detailed understanding of the degradation physics of LEDs can provide the root causes of the unexpected degradation of LEDs so that early detection monitoring can be possible, and their lifetimes can also be enhanced for increased effectiveness in energy saving.

The LED lifetime varies somewhere between 50,000 and 70,000 hours [1, 2, 17]. The LED lifetime is measured by means of luminous flux (lumen) maintenance, which shows the fall of lumen of LEDs over time. The Alliance for Solid-State Illumination Systems and Technologies (ASSIST) has defined LED lifetime based on the time to 50% light output degradation (L50: for the display industry approach) or 70% (L70: for the lighting industry approach) light output degradation at room temperature [18, 19].

However, evaluation of lumen degradation at standard operating conditions for LED lifetime assessment is not feasible as the costs of the test and time are too high. To expedite the reliability test of LEDs, accelerated life tests are used to predict the LEDs lifetime and reliability while maintaining exact failure mechanism between operating and accelerated conditions so that extrapolation can be performed. The most common industrial standards used as qualification tests for LEDs are JEDEC and JEITA standards [20]. These standards include reflow soldering test, thermal shock test, temperature cycle test, moisture resistance cyclic test, high temperature storage test, temperature humidity storage test, low temperature storage test, vibration test, and electro-static discharge test.

The degradations of high-power LEDs are mostly due to three external stresses, namely temperature, humidity and electrical stress [21]. A large number of investigations are done to understand and examine their degradation physics under thermal and electrical stresses. On the contrary, the investigations on the effect of humidity related degradation mechanisms is quite low as reported by Singh & Tan [21]. Previously, it was believed that thermal and electrical stress are the most dominant factors in lumen degradation of LEDs. However, when the applications of LEDs are in outdoor or harsh environmental applications, it was found that instead of the LED chip, the materials of LED encapsulation and molding which is comprised of different silicone polymers degrade in presence of humidity that render major lumen degradation of LEDs in their applications. In this chapter, a detailed study of the dynamics of the degradation physics of silicone polymers in LED encapsulation and molding, and their effect on total lumen degradation of LEDs will be presented.

2. Identification of humidity related failure mechanisms in high-power LEDs

The failure mechanisms induced due to humidity or penetration of moisture into LED packages can be classified mainly into two categories: a) Degradation of the Chip which is further split into semiconductor dice and phosphor; b) degradation of the LED package which includes encapsulation of the LED chip or LED chip + phosphor, LED molding and the die attach with heat sink. A schematic diagram of an LED structure is shown in **Figure 1**.

2.1 Degradation of the chip including phosphor layer

To characterize the chip level degradation in the case of white phosphor LEDs, the computation of the blue to Yellow ratio (BYR) of the output light, and ideality factor (n), Reverse saturation current (I_s) and series resistance between bond wires with chip and bond pads (R_s) are considered [22]. A method of analyzing the n, Is and R_s of the LEDs from the LED forward characteristics curve was developed by Tan et al. which was adopted for all such analysis mentioned hereafter [23]. The analysis of BYR is done in such a way that if the BYR decreases with time, it means the blue light of the LED chip has degraded indicating semiconductor dice degradation. On the other hand, if BYR increases, the yellow light conversion from phosphor decreases

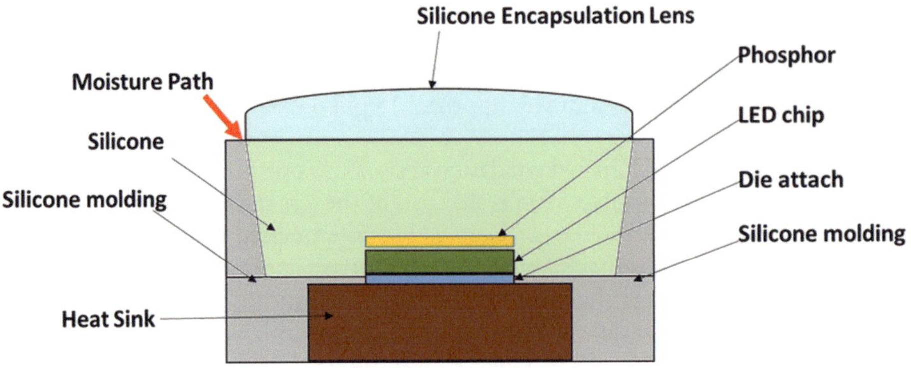

Figure 1.
Schematic of moisture penetration into a typical LED structure. The red arrow represents the possible moisture diffusion paths into the LED package.

indicating the phosphor degradation. This phenomenon is further analyzed by observing the percentage change in the ideality factor (n) where we expect n to go out of the normal range of 2.0–7.0 for chip related degradation [23]. For Blue LEDs without phosphor layer, all the above-mentioned characterizations except BYR are done.

It has been seen that moisture penetration from ambient through the LED package has caused serious lumen degradations. This happened either due to the degradation of the chip or degradation of phosphor or both, and degradation of the silicone encapsulation and molding which will be discussed in the next section. An example of the degradation of the chip due to moisture penetration that reduces the lifetime of the LED chip where its BYR drops to half of its original value at 20% rapidly was reported in Ref.s [22, 24]. The intensity of the emitted blue light degrades over time. Upon further analysis of the cause of failure of the semiconductor dice, it was found that the moisture trapped in the die attach or in the electrical contact pad vaporized and the increase in vapor pressure resulted in the crack of the GaN based die [24, 25].

An example of degradation due to phosphor is the work by Tan et al. [24–26]. The BYR was found to increase for some samples which indicated the degradation of phosphor in LEDs. It was found that moisture trapping inside the LED package has caused the dissolution of phosphor which resulted in the overall lumen degradation.

2.2 Degradation of LED package

It is noteworthy that the humidity test reported until now has a metal reflector below the chip and the effect of moisture on the LED degradation can be identified as drastic [24, 25]. However, with the maturation of chip technology against humidity and the replacement of reflector molding material by the mixture of silicone polymers in high-power LEDs, question on its humidity reliability is re-visited.

To address this question, high temperature-humidity tests at 85°C/85%RH was conducted on OSRAM LEDs according to IPC/JEDEC standard J-STD-020D (2007) [19, 20]. Such test was performed by Tan and Singh [26–28], but they included one more test parameter, namely the drive current of the LEDs, and such reliability tests were termed as moisture-electrical thermal (MET) tests.

The test was performed on two sets of white (with phosphor) and blue (without phosphor) LEDs where one set of each type was kept ON with 350 mA constant current and the other set was turned OFF.

A significant difference in degradation between the ON and OFF sets of each type of LEDs was found as seen in **Figure 2**. The white LEDs in the OFF condition did not experience any lumen degradation which is in agreement with reports from Tan et.al [22, 25, 26] where phosphor acts as a protective layer for the LED. However, in the ON condition, white LEDs experience 33% lumen degradation in a span of 144 hours of testing. On the other hand, a difference of 5% lumen degradation was observed between the ON and OFF blue LEDs and the difference increase to 18% after the tests was concluded for each case. It was also found that the BYR of the fresh and degraded LEDs exhibited insignificant changes which omits the possibility of chip degradation as mentioned in the previous section.

Upon closer examination of **Figure 2**, a sharp degradation of lumen is observed, and it is attributed to the moisture entrapment in the silicone encapsulation of the LEDs. After moisture penetrates through the LED encapsulation, two phenomena are most likely to happen viz. trapping of moisture in silicone encapsulation which causes output light scattering and reduced lumen output; and the diffusion of

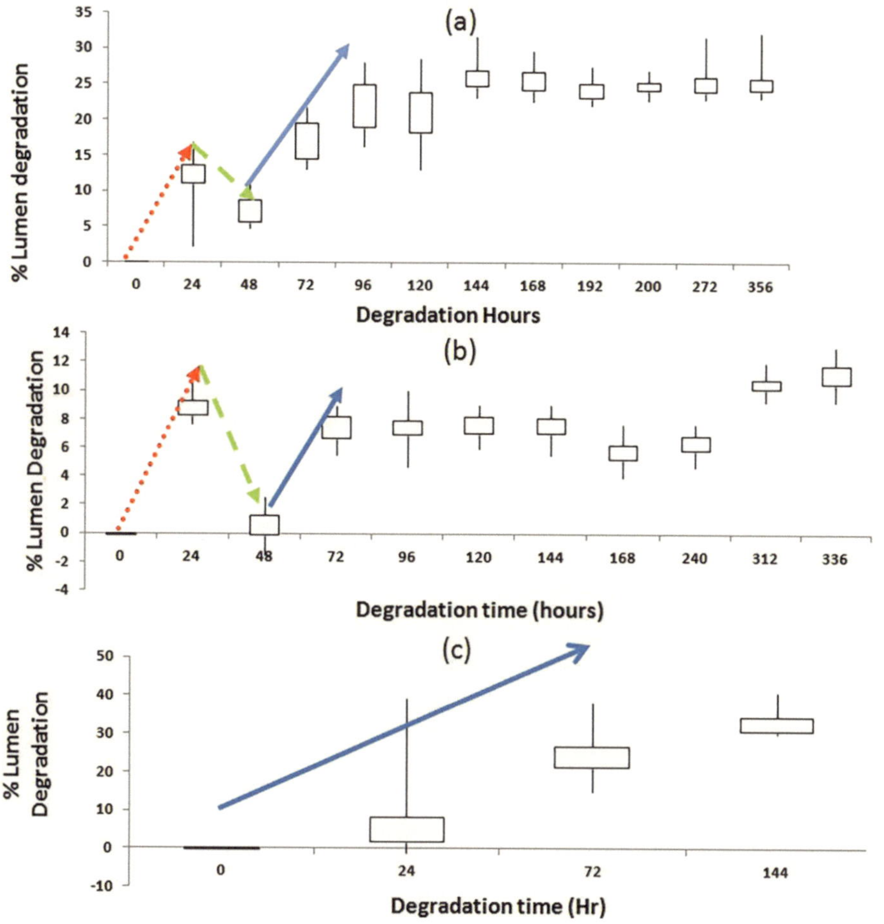

Figure 2.
Percentage lumen degradation vs. test time for blue and white LEDs tested under different operating conditions. (a) Blue LEDs under power-on condition; (b) blue LEDs under power-off condition; and (c) white LEDs under power-on condition. Red line (dotted line) denotes the initial degradation; green line (dash line) indicates the lumen recovery and the blue line (solid line) indicates the final degradation for the LEDs [26].

moisture to the LED chip including phosphor, thereby causing damage in the chip level and resulting in reduced lumen output as discussed in the earlier section. However, such rapid degradation might also have followed by a recovery if the above two damages were not happened. Furthermore, upon further investigations, it was concluded that in order to prevent such moisture penetration, the pore size of the silicone encapsulation must be lower than 50 nm. A detail description of the work can be found in Ref. [22].

3. Failure analysis of humidity reliability of LEDs at OFF state during MET test

Singh et al. identified three stages of degradation mechanisms during the MET test which has been characterized in **Table 2**. The amount of moisture adsorbed by the LED encapsulation and die attach has also been studied by Singh et.al as shown in **Figure 3** which provide good insights in understanding the degradation mechanism of the LEDs. As the degradation mechanisms of LEDs due to moisture for LEDs at ON and OFF conditions are different as evident from the experimental finding, let us discuss the failure mechanism for the LEDs at OFF condition.

It was found that at the beginning of the test, moisture adsorption in the LED encapsulation is much higher than the die attach area and saturates around 132 hours of testing as shown in **Figure 3**. This initial degradation stage is attributed to moisture adsorption in the encapsulant which causes a "cloudiness" or light scattering effect as mentioned in Ref. [25]. The moisture trapped in the silicone encapsulation causes discoloration of the encapsulant material attributing to the "cloudiness" effect which cause lumen degradation. However, for white LEDs the phosphor layer prevents moisture penetration into the die attach region as well as the heat generated from the phosphor during lumen measurements help to evaporate some moisture in the encapsulant. Hence, no significant lumen degradation was observed.

As the test proceeds over 132 hours in OFF conditions, moisture entered the die attach material that caused subsequent die attach delamination which in turn increased the thermal resistance of the die, renders degradation in the overall lumen output of the LEDs. Furthermore, the moisture in the die attach cannot be evaporated during testing [22, 25], and this renders significant lumen degradation. It was found that moisture penetration is 40% higher in blue LEDs as compared to their white counterparts and this excess moisture is housed in the die attach material, therefore when the blue LEDs were tested for 336 hours, die attach delamination has already occurred [26].

For better understanding of the diffusion of moisture in LED degradation, finite element method (FEM) simulation was performed. The three most important

Stages	Blue LED ON	Blue LED OFF	White LED ON	White LED OFF
Initial Degradation	12% degradation	8.5% degradation	Nil	Nil
Recovery	7.5% degradation	1% degradation	Nil	Nil
Final Degradation	30% degradation	12% degradation	33% degradation	Nil

Table 2.
Stages of degradation in LEDs during 85/85 test at ON and OFF conditions [26].

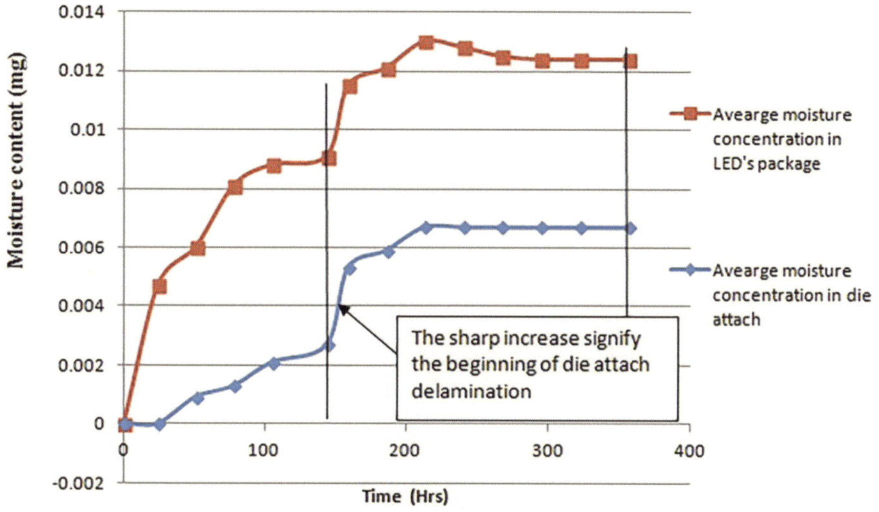

Figure 3.
Amount of moisture penetration into LED package and die attach during testing under OFF condition. The results are computed using finite element analysis as described in the text [26].

parameters considered for such modeling are the moisture diffusivity D which is a function of temperature, moisture saturation concentration C_{sat} and moisture absorption and desorption times. The moisture diffusion in water permeable body is given in Eq. (1) as (**Table 3**) [30].

$$\frac{\partial C}{\partial t} = D(\frac{\partial^2 C}{\partial x^2} + \frac{\partial^2 C}{\partial y^2} + \frac{\partial^2 C}{\partial z^2} \qquad (1)$$

A term W, called normalized concentration or wetness was introduced so that thermal modeling can be applied for moisture modeling as moisture concentration may not be continuous across an interface of two materials, unlike temperature. W is given in Eq. (2) as follows [29]:

$$W = C / C_{sat} \qquad (2)$$

where C is the moisture concentration (kg/m ^{3}) in a material.

It was found that during the absorption cycles, the moisture concentration was highest near the die-attach region and during the desorption cycle, the moisture concentration was high in the silicone than ambient. FEM results also indicated that moisture can reach the die attach under 24 hours of test [31, 32].

Tan et.al [28] used FEM to understand the experimental phenomena of initial rapid light output degradation followed by a recovery. The "recovery" is the result of moisture absorbed by die attach material that suck the moisture from the silicone. Thus the "recovery" of lumen degradation is actually associated with the degradation in the internal structure of the LED package which is not reversible.

Property	Thermal	Moisture
Primary Variable	Temperature, T	Wetness, W
Density	ρ (kg/m^3)	1
Conductivity	k (W/m*^{0}C)	D. C_{sat} (kg/s*m)
Specific heat	c (J/kg*^{0}C)	C_{sat} (kg/m^3)

Table 3.
Correspondence table for thermal/moisture analogy [29].

4. Failure analysis of LEDs at ON state during MET test

From **Table 2**, it is obvious that the LEDs under power-on condition degrade at a faster rate than the one under power-off condition as expected. Since the LEDs at ON state generate heat and are able to reach a temperature of 135°C, the silicone in the encapsulation and molding start to expand due to excess heat. As the material compositions for encapsulation and molding materials (even though both of them are silicone) are different, their difference in thermal co-efficient of expansion led to interfacial void, cracks and delamination as found by Singh et.al [26]. These delaminated interfaces produce gap for moisture to penetrate preferentially into the LED packaging, causing lumen degradation for both white and Blue LEDs.

The Blue LEDs however were found to undergo a recovery phase of light output after 24 hours of the test and the lumen started degrading again after 48 hours. This phenomenon was studied extensively using FEM simulations and experimental verification through confocal scanning acoustic microscopy (c-SAM) [28]. It was found that in the first 24 hours the moisture trapped at the LED encapsulation caused its discoloration and cloudiness which led to a sharp lumen drop of 12%. However, as the LEDs were turned ON, moisture seeping into the die attach occurred thereby exhibiting a recovery stage during the LED test. After 48 hours, delamination in the encapsulation-molding started to occur due to the heat generated by the LED chip. It led to a higher flux of moisture inside the LED package causing constant lumen degradation hereafter.

In case of White LEDs, the phenomenon of reliability recovery was not found as in the case of blue LEDs, and a total degradation of 33% occurred continuously till 144 hours. To further understand the root cause of this degradation, optical microscope imaging was used to observe changes in LEDs after every 24 hours of testing. The optical images revealed heavy discoloration of the LEDs which was suspected to be in the silicone encapsulation and molding part of the LEDs as shown in **Figure 4** [33].

The discoloration in the molding part and encapsulation of the white LEDs were found to be more severe as compared to blue LEDs due possibly to the excess heat generated by the phosphor layer during the light conversion process [33]. Destructive analysis of the LED encapsulation and molding parts revealed that cracks were observed in both components. However, the cause of such cracks in both encapsulation and molding parts of the LED package were different as reported by Singh et.al [33, 34].

4.1 Analysis of encapsulant degradations

The surface of the fresh LEDs was found to be smooth, while after degradation, cracks started to appear and flakes or tube like structures were observed on the encapsulation of the blue LEDs. The depth of cracks on the encapsulant was quite minimal where average depth was around 3 μm for blue LEDs and that for white degraded LED encapsulant, it is almost negligible and not able to measure using optical microscope [33].

Scanning Electron Microscopy (SEM) was employed to further investigate the surface of the blue and white degraded LED's encapsulant and compared to a fresh LED encapsulant. SEM images also found cracks on the degraded encapsulants which was consistent with optical imaging results. The flake like structures were more evident on the blue degraded LEDs surface as compared to the white degraded LEDs. Both SEM and optical micrographs shows that the crack density and crack depth are higher in blue LED's encapsulant than the white LED [33].

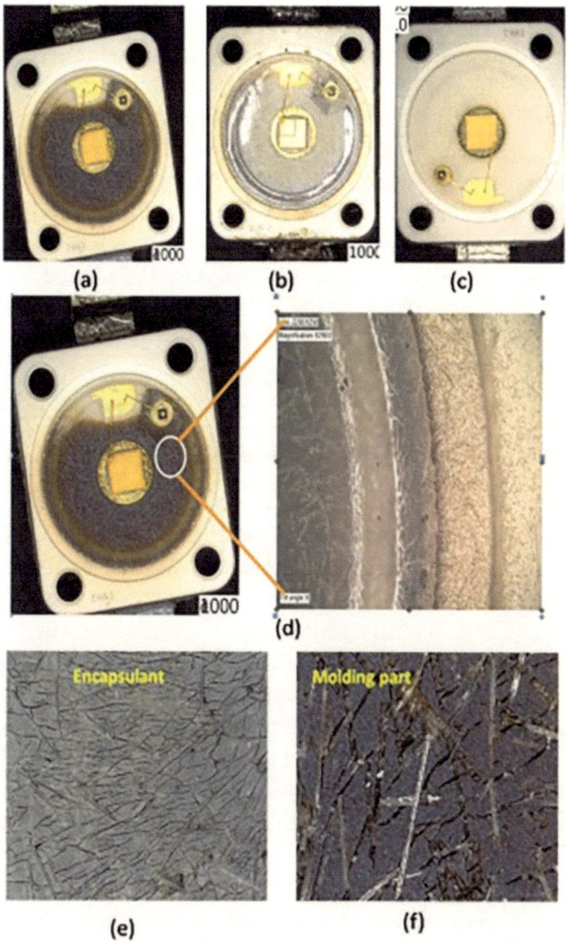

Figure 4.
Optical micrograph of silicone discoloration in (a) white LEDs tested for 144 hours, (b) blue LEDs tested for 356 hours respectively when compared to (c) fresh LEDs. (d) a closer look of the top view of the degraded white LED. Encapsulant and molding parts are also detached from one another for detail examination. Optical micrographs at 500 X magnification of (e) encapsulant and (f) molding part of degraded white LEDs tested under 85°C/85% RH and ON condition for 144 hours [33]. (reproduced with permission from [33]).

However, if the total rise in temperature of both blue and white LEDs are compared when LEDs are powered on at the same drive current, it can be easily found that the total temperature of the white LEDs are more. The higher temperature of white LEDs is attributed to mainly two reasons; the junction temperature of the LED junction in the die and the heat generated when phosphor layer is activated for conversion of blue to yellow light. Therefore, it is expected that, in comparison to the percentage lumen degradation of both blue and white LEDs, and taking into account the heat generation in white LEDs, the crack in the encapsulation of the white LEDs should be more than blue LEDs. However, this was not the case, and as reported above, the encapsulants of blue LEDs experienced severe cracks than the white LEDs.

Energy dispersive system (EDS) was employed to examine the crack surface of the degraded and fresh encapsulants to provide a better understanding of the nature of crack on the encapsulants. EDS results showed that the percentage increase in oxygen is higher in the case of the blue degraded LED encapsulant as compared to that of the white LEDs. A higher percentage in oxygen element could indicate that the surface of the material has undergone oxidation which can be through a series of chemical reactions. To narrow down the type of chemical reaction which has occurred in the encapsulant surface, FTIR spectroscopic analysis was employed. FTIR analysis indicated that the degraded encapsulant of both the LEDs contained –OH peaks which indicates degradation by hydrolysis. However, the OH peaks concentration was higher in case of blue LEDs than its white counterpart. This means that moisture accumulation in white LED encapsulants was lower compared to that in blue LEDs, and thus less severe hydrolysis of the white LEDs than blue. The moisture concentration of the white LEDs was lower due to excess heat generated by the phosphor layer that drives away the moisture from the surface of the LED encapsulant [33]. The degradation mechanism related to cracks in encapsulation was thus found to be hydrolysis of the silicone polymers of the LED encapsulant [33, 34].

4.2 Analysis of molding part degradations

The degradation mechanisms of the molding part of the LEDs have not been extensively investigated, to the best knowledge of the authors, and we now present detailed investigation of the white LED molding degradation in this section. It was found that alongside with the yellowing of the molding part, cracks also started to develop as the test progressed from 45 to 150 hours. Optical micrograph investigation of the crack depths revealed that the average crack depth was found at 105 hours of testing corresponding to 25–27 μm. When the test progressed towards 150 hours, depths began to decrease by an average of 10 μm. This indicates that different types of chemical changes in the silicone molding must have taken place to observe such changes [34].

This hypothesis was later confirmed by the reports from FTIR analysis where it revealed two sets of exothermic and endothermic reactions in the silicone molding part. It was found that in the initial phases of the test (0–45 hours), the moisture penetration towards the molding part of the LEDs caused the hydrolysis of the silicone polymers which is an endothermic reaction [34–36]. The endothermicity of hydrolysis was further confirmed by the drop in temperature in the LED packages as reported by Singh et al. [34]. The endothermic nature of the hydrolysis reaction was further confirmed by Shabir et.al using density functional theory (DFT) calculations where it was found that 460 kJ/mol of energy was absorbed during hydrolysis of the silicone polymers under study [35]. Hydrolysis reaction was immediately followed by condensation of the silicone polymers which is an exothermic reaction where a raise in temperature was also observed by Singh et.al and others [34, 37]. Shabir et al also verified that condensation is exothermic as DFT calculations show 545 kJ/mol of energy release during condensation reaction [35].

As seen from **Figure 5**, it is evident that in the initial 48–72 hours of test, discoloration of the molding part was prominent, and the development of cracks was negligible. DFT calculations on the hydrolysis and condensation of silicone polymers found that condensation reaction is highly spontaneous as compared to hydrolysis during this time. Also, due to the exothermic nature of condensation reaction, the energy released from condensation is continuously re-absorbed by the moisture molecules in the outer part of the molding, thereby increasing the rate of hydrolysis exponentially after some point of time [35]. It was hence found that the sudden increase in lumen degradation was due to simultaneous hydrolysis and condensation reactions as reported by Shabir et al. [35].

As the temperature rises in the sites of the molding part where condensation has occurred, the decrease in Si-O peak and the increase in H_2O peaks indicates that a different chemical reaction occurs where the involvement of H_2O is reduced. It was found that silicone network crosslinking can occur producing silicone ring-like structure due to thermal oxidation at temperatures of 0-200°C which is exothermic in nature [34, 38, 39]. The occurrence of thermal oxidation in those sites after condensation at 48 hours of test is believed to lead to crack depth formation which is the dominant degradation mechanism until 105 hours.

The phenomenon of crack recovery after 105 hours was seen. Silicone loses its elasticity after degradation especially in harsh environmental conditions as observed by many researchers in their studies [40]. Thus, possibility of molding part's crack

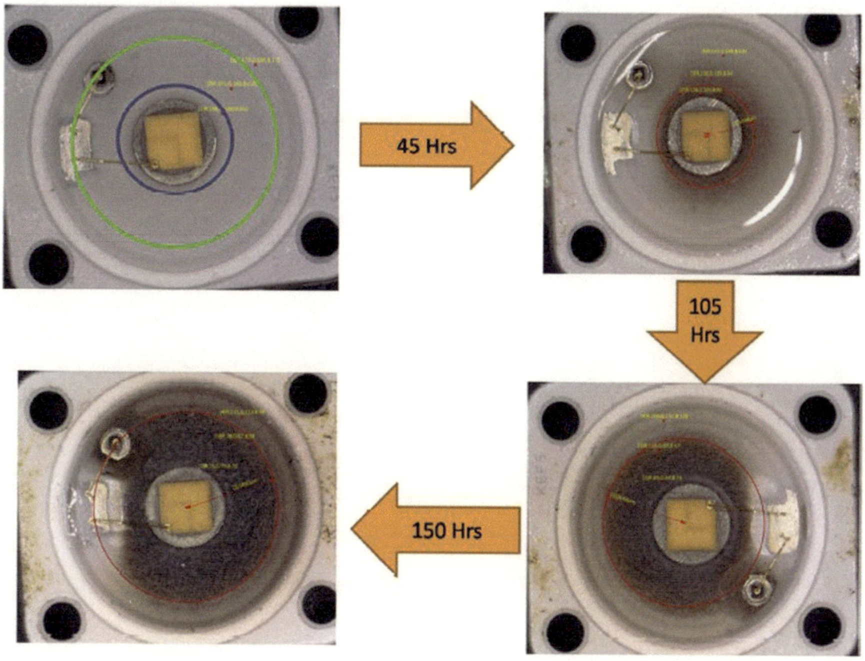

Figure 5.
Optical micrographs of white LEDs with varying test time intervals. The blue colored circle in the fresh sample represents the inner molding part close to LED chip periphery and the green colored circle represents the outer molding part which is far from the LED chip [34]. (reproduced with permission from [34]).

recovery due to its elastic nature is not valid. Another possibility is due to the diffusion of low molecular weight species such as Si-O (also known as Si oligomers or silica nano-fillers) from the interior of silicone to its surface as it could be invoked by the high temperature of the silicone due to thermal oxidation. The Si oligomers cover the cracks area, renders an apparent crack depth recovery [39]. This is also observed as an increase in Si-O peak in FTIR results as seen in **Figure 6**. EDS results also supports this theory showing an increase in the Si atoms after 45 h of testing appear to confirm the presence of Silicon oligomers. In other words, as cracks are generated due to thermal oxidation, Silicon oligomers can now diffuse out from the crack surfaces, filling the cracks and thus the cracks depth decrease.

As the LED package degrades, the temperature of the device also increases as shown in Ref. [34]. At higher temperature, thermal aging of silicone occurs as evidenced by the increase in Si-CH_3 molecules in the FTIR results. Although thermal aging is an endothermic reaction as thermal energy is required to break the chemical cross-linking bonds in the silicone structure, the temperature of the LED does not decrease when thermal aging step in. This could be due to the degradation at either LED chip site or at the die attach as observed by many [24, 26] which can result in additional temperature rise and thus offset the decrease in temperature due to thermal aging reaction. In fact, this additional temperature rises at the LED chip or die attach sites will lead to rapid diffusion of oligomers in the region closer to the LED chip than the area far away from the chip region, and this is indeed observed by Singh et al. [34] where higher amount of crack depth recovery is observed for inner region cracks when compared with outer region cracks from 105 to 150 h duration.

However, the bond dissociation energy of Si-O-Si bond varies between 12 and 162 KJ/mol as reported in many literatures [41–46]. So, if we consider the energy of ambient temperature of 85°C, it only corresponds to 2.97 kJ/mol. If the LED temperature is also added with the LEDs powered ON, it only corresponds to 3.89–5.7 kJ/mol. It is therefore clear that the energy from temperature cannot be the sole driving force for such kinds of degradations.

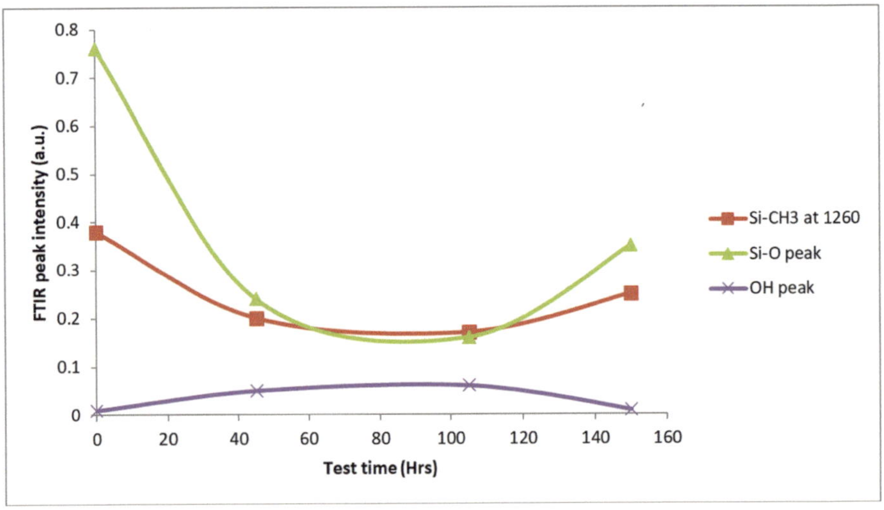

Figure 6.
Overall peaks intensity variation for different samples with varying test time [34]. (reproduced with permission from [34]).

To provide detailed understanding of the driving force of the degradation mechanisms as mentioned above, Shabir et.al performed ab-initio calculations on the hydrolysis and condensation of silicone polymers [35]. They found that the activation energy E_a of hydrolysis of silicone was 123.2 kJ/mol while that for condensation it was −6.9 kJ/mol. The E_a of hydrolysis computed in this study was found to be in excellent agreement with the E_a of hydrolysis of silicone polymers present in the LED molding experimentally. The driving force of such reactions was found dominantly from the mixture of blue and yellow light emitted from the LEDs during the test [35]. This also agrees excellently with the test reports where white LEDs in OFF conditions exhibited no lumen degradations as mentioned earlier [26, 28].

As LEDs in outdoor applications are designed to perform for longer periods of time, lifetime prediction of LEDs is done with accelerated stress levels of temperature and humidity know as accelerated life tests (ALTs), and then lifetime at such conditions are extrapolated to estimate their lifetimes under normal operating conditions. The underlying idea of such extrapolation is that the devices under test should follow the same degradation mechanisms in both accelerated and normal operating conditions. The common acceleration models used for reliability test of LEDs are the Peck models which includes the extrapolation of test data from 85%/85°C to normal operating conditions and the Arrhenius equation is included for temperature effect therein. However, it was found by Tan et.al that the degradations are completely different for humidity level varying between 70% RH, 85% RH and 95% RH [24]. This renders the inapplicability of the Peck model for extrapolation of test data to normal operating condition for LEDs. Peck model also does not consider the effect of light energy on the degradation of LEDs. On the temperature extrapolation, Arrhenius equation is commonly used, and the assumption of constant activation energy is made, which indicate that the degradation mechanism is invariant as it is the basis of extrapolation as mentioned earlier. However, our study also found that this is not true. Hence there is a need for a new model which quantitatively covers all the factors for silicone degradation, and this model was developed by Shabir et.al from detailed understanding of DFT simulations and experimental data as well as the of physics of degradation from various reports [35]. The new model related the growing of the area of discoloration A_d with time (t) is given in Eq. (3) below.

$$A_d(t) = A_0 + A_2 e^{A_1 t} \tag{3}$$

where

$$A_1 = \frac{\sum_{i=1}^{2} \beta_i P_i}{\alpha}; \tag{4}$$

P_1 represents the power density of the blue light and P_2 represents the power density of the yellow light. The corresponding β represents the absorption coefficient of the different lights by the discolored region, respectively, which was assumed to be constant in this work. As the discolored region is dark brown in color, it was considered that all the blue light is absorbed, i.e., $\beta_1 = 1$.

$\alpha = \left(k_1 E_a - \frac{k_B T}{V} - \alpha_{condensation} \beta_3 \right) \delta$; Here, δ is the effective penetration depth of light. At a given ambient temperature, α is approximately a constant, E_a is the activation energy, k_B is Boltzmann constant, T is temperature and V is volume of the silicone molding the meaning of other parameters can be found in Ref. [35].

A_2 is related to the quality of the silicone. Specifically, if the silicone has lager pore size or trapped particles that can absorb light, the value of A_2 will be larger; $A_0 = -A_2$.

The model fits very well to the experimental data at different light intensities and drive currents as shown in **Figure 7**. It also predicts no degradation if no light is present which was the case of white LEDs with power OFF during testing as reported by Singh et.al [26]. The model allows us to predict LED molding discoloration in the future and simultaneously quantify the quality of the silicone material of the LED molding, which is a factor that will affect the variation of the degradation time of LEDs.

5. Conclusion

Reliability tests and causes of degradation have become important criteria of focus for large scale product manufacturing companies. Proper reliability test and analysis of the related cause of degradation shall help increase the product lifetime and market value. Hence, a descriptive review on the various degradation analysis of LEDs is presented here.

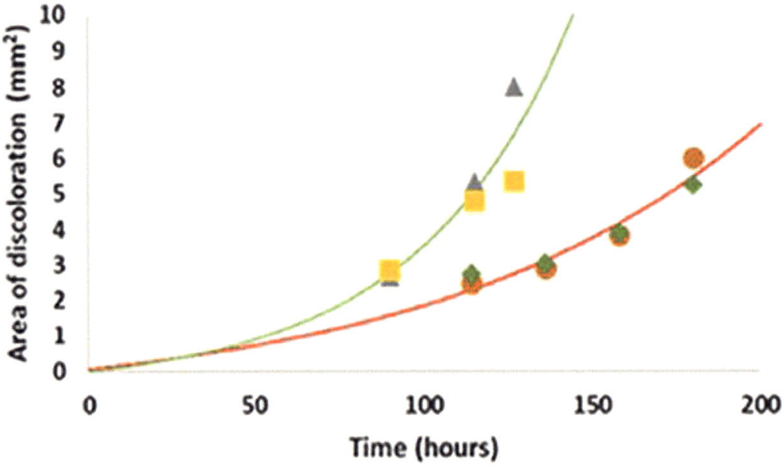

Figure 7.
Model fitting with experimental data for 350 mA and 300 mA drive currents [35]. (reproduced with permission from [35]).

After detailed analysis using experimental and DFT techniques, it was found that the degradation of the encapsulant and molding part material, both make of silicone, are the primary sources of lumen degradation. The inapplicability of Peck's model was also found as it failed to factor-in the primary driving force of silicone degradation which is the light emitted from the LED chip itself. A new model for prediction of silicone degradation was hence developed which determines the degradation of silicone as a function of time, considering all its driving forces. This model allows prediction of the quality of silicone used for LED packaging which in turn can improve the LED reliability. It can also help to predict the rate of discoloration of the LED which relate to its lumen degradation.

Author details

Abdul Shabir[1,2] and Cher Ming Tan[1,2*]

1 Center for Reliability Science and Technology, Chang Gung University, Taiwan

2 Department of Electronic Engineering, Chang Gung University, Taiwan

*Address all correspondence to: cmtan@cgu.edu.tw

References

[1] Farsakoglu OF, Yusuf H, Kilis H, Üniversitesi A, Farsakoglu OF, Hasirci HY. Energy optimization of low power LED drivers in indoor lighting Comprehensive Design of Stirling Engine Based Solar Dish Power Plant with solar tracking system view project DETERMINATION OF THE EFFECT ON THE ENVIRONMENT AND ENERGY EFFICIENCY OF LED AND CO. Artice Journal of Optoelectronics Advance Materials. 2015, Accessed: July 03, 2022;**17**(6):816-821 [Online]. Available: https://www.researchgate.net/publication/283744399

[2] De Almeida A, Santos B, Bertoldi P, Quicheron M. Solid state lighting review - potential and challenges in Europe. Renewable and Sustainable Energy Reviews. 2014;**34**:30-48. DOI: 10.1016/j.rser.2014.02.029

[3] Guan N, Dai X, Babichev AV, Julien FH, Tchernycheva M. Flexible inorganic light emitting diodes based on semiconductor nanowires. Chemical Science. 2017;**8**(12):7904-7911. DOI: 10.1039/c7sc02573d

[4] Krames MR et al. Status and future of high-power light-emitting diodes for solid-state lighting. IEEE/OSA Journal of Display Technology. 2007;**3**(2):160-175. DOI: 10.1109/JDT.2007.895339

[5] IEEE Xplore Full-Text PDF https://ieeexplore.ieee.org/stamp/stamp.jsp?tp=&arnumber=999186 [Accessed: June 21, 2022].

[6] High Power LEDs – Technology Status and Market Applications - Steranka - 2002 - physica status solidi (a) - Wiley Online Library https://onlinelibrary.wiley.com/doi/abs/10.1002/1521-396X(200212)194:2%3C380::AID-PSSA380%3E3.0.CO;2-N [Accessed: June 21, 2022]

[7] Schubert EF, Kim JK, Luo H, Xi JQ. Solid-state lighting—A benevolent technology. Reports on Progress in Physics. 2006;**69**(12):3069. DOI: 10.1088/0034-4885/69/12/R01

[8] Schubert EF, Kim JK. Solid-state light sources getting smart. Science (80-.). 2005;**308**(5726):1274. DOI: 10.1126/science.1108712

[9] Bergh A, Craford G, Duggal A, Haitz R. The promise and challenge of solid-state lighting. Physics Today. 2001;**54**(12):42. DOI: 10.1063/1.1445547

[10] Aoyama Y, Toshiaki Y. An LED module array system designed for streetlight use. 2008 IEEE Energy 2030 Conference. IEEE; 2008. DOI: 10.1109/ENERGY.2008.4780996

[11] New Solid State Technologies and Light Emission Diodes as a mean of Control and Lighting Source Applicable to Explosion Proof Equipment, with the Scope to Reduce Maintenance, to Limit the risk of bad Maintenance and to Expand the Plants' Life | IEEE Conference Publication | IEEE Xplore https://ieeexplore.ieee.org/abstract/document/5164883 [Accessed: June 21, 2022]

[12] Moreira MC, Prado R, Campos A. Application of high brightness LEDs in the human tissue and its therapeutic response. In: Applied Biomedical Engineering. London, UK: IntechOpen; 2011

[13] China's LED Lighting Fixtures are Frequently Recalled, and Quality Problems Constrain International

Competitiveness https://www.1ledlight.com/chinas-led-lighting-fixtures-are-frequently-recalled-and-quality-problems-constrain-international-competitiveness.html

[14] Halco recalls LED bulbs due to risk of Injury and Burn Hazards. 2014 https://www.cpsc.gov/zhT-CN/Recalls/2014/halco-recalls-led-bulbs

[15] Cooper Lighting Recalls Solar/Battery Powered Light Fixtures Due to Fire Hazard https://www.cpsc.gov/Recalls/2018/cooper-lighting-recalls-solarbattery-powered-light-fixtures-due-to-fire-hazard

[16] Ferretti C. Detroit's LED streetlights going dark after a few years; 2019. Available from: https://www.detroitnews.com/story/news/local/detroit-city/2019/05/07/detroits-led-streetlights-going-dark-after-few-years/3650465002/

[17] Chang MH, Das D, Varde PV, Pecht M. Light emitting diodes reliability review. Microelectronics and Reliability. 2012;**52**(5):762-782. DOI: 10.1016/j.microrel.2011.07.063

[18] Kaufman JE, editor. IES Lighting Handbook: Reference Volume. Illuminating Engineering Society of North America; 1984. ISBN: 978-0-87995-010-1

[19] Ohno Y. Optical Technology Division IESNA standards on LED and SSL: LM-79, LM-80, and future standards. Available from: https://cormusa.org/wp-content/uploads/2018/04/CORM_2009_-_IESNA_Standards_on_LED_and_SSL_LM79LM80_and_Future_Standards_CORM_2009_Y_Ohno.pdf

[20] J-STD-20. Joint IPC/JEDEC Standard for Moisture/Reflow Sensitivity Classification for Nonhermetic Surface-Mount Devices. no. E [Online]. 2014. Available: https://www.jedec.org/standards-documents/docs/j-std-020e [Accessed: July 03, 2022]

[21] Singh P, Tan CM. A review on the humidity reliability of high power white light LEDs. Microelectronics and Reliability. 2016;**61**:129-139. DOI: 10.1016/j.microrel.2015.12.002

[22] Tan CM, Eric Chen BK, Xu G, Liu Y. Analysis of humidity effects on the degradation of high-power white LEDs. Microelectronics and Reliability. 2009;**49**(9-11):1226-1230. DOI: 10.1016/j.microrel.2009.07.005

[23] Tan CM, Gan Z, Ho WF, Chen S, Liu R. Determination of the dice forward I–V characteristics of a power diode from a packaged device and its applications. Microelectronics and Reliability. 2005;**45**(1):179-184. DOI: 10.1016/J.MICROREL.2004.05.004

[24] Tan CM, Singh P. Time evolution degradation physics in high power white LEDs under high temperature-humidity conditions. IEEE Transactions on Device and Materials Reliability. 2014;**14**(2):742-750. DOI: 10.1109/TDMR.2014.2318725

[25] Tan CM, Chen BK, Li X, Chen SJ. Rapid light output degradation of GaN-based packaged LED in the early stage of humidity test. IEEE Transactions on Device and Materials Reliability. 2012;**12**(1):44-48. DOI: 10.1109/TDMR.2011.2173346

[26] Singh P, Tan CM. Degradation physics of high power LEDs in outdoor Environment and the role of phosphor in the degradation process. Scientific Reports. 2016;**6**(1):1-13. DOI: 10.1038/srep24052

[27] Tan CM and Singh P. Extrapolation of lifetime of high power LEDs under

temperature-humidity conditions. In: 2014 IEEE International Conference on Electron Devices and Solid-State Circuits. IEEE; 2014. DOI: 10.1109/EDSSC.2014.7061079

[28] Singh P, Tan CM, Chang LB. Early degradation of high power packaged LEDs under humid conditions and its recovery — Myth of reliability rejuvenation. Microelectronics and Reliability. 2016;**61**:145-153. DOI:10.1016/J.MICROREL.2015.12.036

[29] Wong EH, Teo YC, Lim TB. Moisture diffusion and vapour pressure modeling of IC packaging. Proceedings - 48th Electronic Components and Technology Conference. 1998;**Part F133492**:1372-1378. DOI: 10.1109/ECTC.1998.678922

[30] Ma X et al. Moisture diffusion model verification of packaging materials. In: 2008 International Conference on Electronic Packaging Technology & High Density Packaging. IEEE; 2008. DOI: 10.1109/ICEPT.2008.4607066

[31] Wu B, Luo X, Liu S. Effect mechanism of moisture diffusion on LED reliability. In: 3rd Electronics System Integration Technology Conference ESTC. IEEE; 2010. DOI: 10.1109/ESTC.2010.5642833

[32] Tan L, Li J, Wang K, Liu S. Effects of defects on the thermal and optical performance of high-brightness light-emitting diodes. IEEE Transactions on Electronics Packaging Manufacturing. 2009;**32**(4):233-240. DOI: 10.1109/TEPM.2009.2027893

[33] Singh P, Tan CM. Uncover the degradation science of silicone under the combined temperature and humidity conditions. IEEE Access. 2017;**6**:1302-1311. DOI: 10.1109/ACCESS.2017.2778289

[34] Singh P, Tan CM. Time evolution of packaged LED lamp degradation in outdoor applications. Optical Materials (Amst). 2018;**86**(July):148-154. DOI: 10.1016/j.optmat.2018.10.009

[35] Shabir A, Tan CM. Impact of visible light and humidity on the stability of high-power light emitting diode packaging material. Journal of Applied Physics. 2021;**130**(8):083101. DOI: 10.1063/5.0059515

[36] Х А Б А Р Л А Р Ы ИЗВЕСТИЯ НАЦИОНАЛЬНОЙ АКАДЕМИИ НАУК РЕСПУБЛИКИ КАЗАХСТАН ХИМИЯ ЖӘНЕ ТЕХНОЛОГИЯ СЕРИЯСЫ. СЕРИЯ ХИМИИ И ТЕХНОЛОГИИ: SERIES CHEMISTRY AND TECHNOLOGY 6 (420) [Online]. Available: http://nauka-nanrk.kz [Accessed: July 09, 2022]

[37] Pérez-Quintanilla D, Del Hierro I, Fajardo M, Sierra I. Adsorption of cadmium(II) from aqueous media onto a mesoporous silica chemically modified with 2-mercaptopyrimidine. Journal of Materials Chemistry. 2006;**16**(18):1757-1764. DOI: 10.1039/b518157g

[38] Song L, He Q, Hu Y, Chen H, Liu L. Study on thermal degradation and combustion behaviors of PC/POSS hybrids. Polymer Degradation and Stability. 2008;**93**(3):627-639. DOI: 10.1016/J.POLYMDEGRADSTAB.2008.01.014

[39] Sun P, Horton JH. Perfluorinated poly(dimethylsiloxane) via the covalent attachment of perfluoroalkylsilanes on the oxidized surface: Effects on zeta-potential values. Applied Surface Science. 2013;**271**:344-351. DOI: 10.1016/J.APSUSC.2013.01.199

[40] Bao YW, Wang W, Zhou YC. Investigation of the relationship between elastic modulus and hardness based on depth-sensing indentation

measurements. Acta Materialia. 2004;**52**(18):5397-5404. DOI: 10.1016/j.actamat.2004.08.002

[41] Hühn C, Erlebach A, Mey D, Wondraczek L, Sierka M. Ab initio energetics of Si–O bond cleavage. Journal of Computational Chemistry. 2017;**38**(27):2349-2353. DOI: 10.1002/jcc.24892

[42] Atkinson BK. A fracture mechanics study of subcritical tensile cracking of quartz in wet environments. Pure and Applied Geophysics PAGEOPH. 1979;**117**(5):1011-1024. DOI: 10.1007/BF00876082

[43] Gerberich WW, Stout M. Discussion of thermally activated approaches to glass fracture. Journal of the American Ceramic Society. 1976;**59**(5-6):222-225. DOI: 10.1111/j.1151-2916.1976.tb10938.x

[44] Taylor NW. Mechanism of fracture of glass and similar brittle solids. Journal of Applied Physics. 1947;**18**(11):943-955. DOI: 10.1063/1.1697579

[45] Stuart DA, Anderson OL. Dependence of ultimate strength of glass under constant load on temperature, ambient atmosphere, and time. Journal of the American Ceramic Society. 1953;**36**(12):416-424. DOI: 10.1111/j.1151-2916.1953.tb12831.x

[46] Cypryk M, Apeloig Y. Mechanism of the acid-catalyzed Si-O bond cleavage in siloxanes and siloxanols. A theoretical study. Organometallics. 2002;**21**(11):2165-2175. DOI: 10.1021/om011055s

Section 3

GaN LEDs

Chapter 5

Recent Advancements in GaN LED Technology

Thamer A. Tabbakh, Deepak Anandan, Michael J. Sheldon, Prashant Tyagi and Ahmad Alfaifi

Abstract

Gallium nitride (GaN)-based solid state lighting technology has revolutionized the semiconductor industry. The GaN technology has played a crucial role in reducing world energy demand as well as reducing the carbon footprint. As per the reports, the global demand for lighting has reduced around 13% of total energy consumption in 2018. The Department of Energy (USA) has estimated that bright white LED source could reduce their energy consumption for lighting by 29% by 2025. Most of the GaN LEDs are grown in c-direction, and this direction gives high growth rate and good crystal integrity. On the other hand, the *c*-plane growth induces piezoelectric polarization, which reduces the overall efficiency of LEDs since the last decade researchers round the globe working on III-N material to improve the existing technology and to push the limit of III-V domain. Now, the non-polar and semi-polar grown LEDs are under investigation for improved efficiency. With the recent development, the GaN is not only limited to lighting, but latest innovations also led the development of micro-LEDs, lasers projection and point source. These developments have pushed GaN into the realm of display technology. The miniaturization of the GaN-based micro-LED and integration of GaN on silicon driving the application into fast response photonic integrated circuits (ICs). Most of the recent advancements in GaN LED field would be discussed in detail.

Keywords: GaN, polar, non-polar, polarization, GaNLEDs, GaN micro LEDs

1. Introduction

Group III-nitrides (GaN, AlN, and InN) and their alloys have been considered the most promising semiconductor materials for various optoelectronic applications due to their excellent physical properties and stability in harsh environmental conditions [1–3]. Today, III-nitrides-based light-emitting diodes (LEDs) are widely used for solid-state lighting (SSL) applications all over the world because of their high efficiency, low power consumption, and longer lifetime than fluorescent and incandescent bulbs [4, 5]. Specifically, white light LEDs are a more promising low-power light resource to replace conventional fluorescent, as shown in **Figure 1**. Along with LEDs, III-nitride-based laser diodes (LDs), high-power electronics, photodetectors, etc., are other extended optoelectronic applications that are also demonstrated [7, 8]. However,

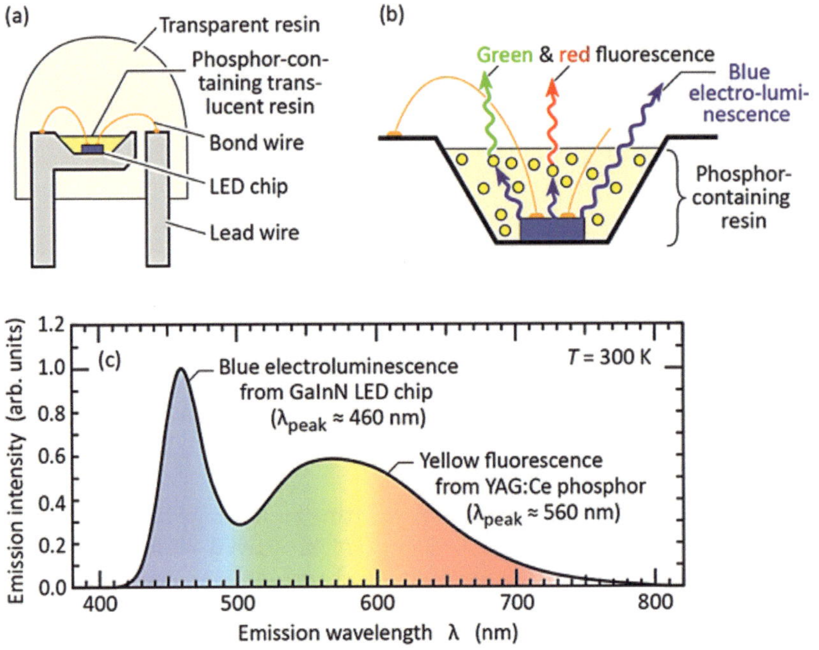

Figure 1.
(a) Structure of white LED, (b) YAG-ce phosphor used to generate white LED by blue emission; and (c) emission spectra of first white LED (covers blue – Yellow- red) [4], where the typical distance between +ve and -ve lead wire is 0.3 mm for 5 mm LED [6].

developing III-nitride-based devices is not straightforward due to various difficulties and challenges. Among these, the preparation of high-quality and large-area single-crystalline bulk III-nitride semiconductors which has been one of the significant challenges over the last few decades because of the poor solubility of nitrogen gas in III-metals (In, Ga and Al) [9–11]. For example, GaN single crystal growth requires a high growth temperature (2220°C) and high growth pressure (6 GPa) [9]. The bulk single-crystal GaN growth cannot be employed in the standard Czochralski or Bridgman techniques, generally used to prepare bulk Si and GaAs substrates [11]. However, to take advantage of III-nitrides material physical properties and their chemical stability, III-nitrides material is hetero-epitaxially grown on various substrates [1, 3].

In semiconductor epitaxy, the most favorable approach is homoepitaxy in which same substrate is used for the growth of targeted semiconductor material. But this approach is not possible for III-nitride based material due to the unavailability of bulk GaN substrates. Hence, hetroepitaxy route is chosen for the growth of GaN based material. Hetroepitaxy means growth of material on foreign substrate. For example, most of the GaN devices are grown on sapphire which is highly lattice mismatched with GaN. Due to heteroepitaxial growth process the grown epilayers suffer from high level of in-plane strain and structural defects like dislocation, stacking faults, etc. Generally, nitridation of sapphire is performed before GaN growth as pre-nitridation step. This pre-nitridation helps in formation of an intermediate AlN layer which compensate for plane strain to some extent [12–15]. Because of the said reason, large area singe- crystalline bulk III-nitride semiconductor fabrication is challenging. Over the last few decades, research has been done to improve the structural, optical, and electronic properties of GaN devices.

1.1 Breakthrough: *P*-type doping

Today, scientific advancement brings tremendous progress in improving the crystalline quality of III-nitrides semiconductors, which enable the development of highly efficient and durable III-nitride-based optoelectronic devices for commercial applications [4, 5, 7, 8]. The GaN material was first time discovered by Johnson *et al.* in the year of 1932. This work obtained polycrystalline GaN by passing ammonia (NH_3) gas over hot gallium metal at 900–1000°C temperature conditions [16]. However, the epitaxial growth of GaN film on sapphire (0001) was presented in 1969s by Maruska and Tietjen using the hydride vapor phase epitaxy (HVPE) growth technique [17]. Thereafter, Manasevit *et al.* reported the growth of epitaxial GaN film on a sapphire substrate using metalorganic vapor phase epitaxy (MOVPE) [18]. However, these hetero-epitaxial grown GaN films on sapphire substrates have poor crystal quality and high-density *n-type* carriers ($>10^{18}$ cm^{-3}), which makes it challenging to achieve *p-type* doping into the GaN films [19, 20]. Yoshida *et al.* addressed that the molecular beam epitaxy (MBE) growth of GaN film on AlN-coated sapphire has shown better electrical and optical properties than the directly grown GaN on the sapphire substrate [21]. Amano *et al.* reported that the thin AlN buffer layer on the sapphire substrate improves the epitaxial GaN film crystalline quality [22]. Nakamura *et al.* also noted the growth of enhanced crystalline quality GaN on the low-temperature GaN buffer layer on the sapphire substrate before the primary GaN film growth [23].

The major development of III-nitride optoelectronic devices was initiated by Amano *et al.* after successfully *p-type* (Mg-acceptor) doping into the GaN film by using a low-energy electron beam irradiation method, which has opened a way to develop the p-n junction-based LEDs [24]. However, it was noticed that the Mg acceptor doped GaN film has a high electrical resistivity due to the formation of acceptor-H complexes [3, 20]. Nakamura *et al.* have shown that the p-GaN film hole carrier activation was achieved by a simple thermal annealing process in the N_2 gas environment [25]. The high temperature thermally annealed p-GaN film has shown a high hole carrier concentration of $\sim 3 \times 10^{17}$ cm^{-3} with mobility of 10 cm^2/Vs and low contact electrical resistivity of 2 Ω.cm as shown in **Figure 2** [25]. Furthermore, the growth of high crystalline InGaN film on GaN film by using metalorganic chemical vapor deposition (MOCVD) has a strong band-edge emission peak at room temperature photoluminescence (PL), which is another breakthrough for developing LEDs devices [27]. In 1993, Nakamura *et al.* successfully reported the first GaN-based blue LED with a light output power of 125 μW and external quantum efficiency (EQE) of 0.22% [26]. Later, the significant development progress in III-nitride semiconductor research enabled highly efficient SSL LED devices (**Figure 3**).

1.2 III-N alloy for wide spectral emission

III-N-based materials are the leading contender to fill the wavelength from 500 to 600 nm. Alloys of AlGaInN/GaN cover a broad spectrum of deep ultraviolet to near-infrared (AlN-6.2 eV, GaN-3.4 eV, and InN-0.65 eV) by easing the severe bottleneck of low-output efficiency by other semiconductor emitters such as ZnSe as shown in **Figure 1**.

Around the 1990s, Nakamura *et al.* successfully developed high-quality GaN epi-film and made efficient p-type GaN film using high-temperature annealing to realize high-efficiency blue-light LED. Taking advantage of the momentum, researchers

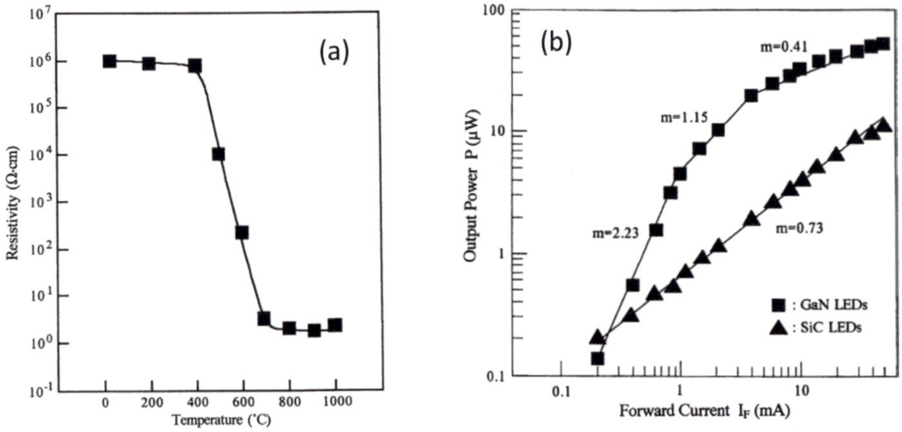

Figure 2.
Resistivity of Mg-doped GaN films for different annealing temperatures. (b) Output power comparison between commercially available SiC LEDs and p-n junction GaN LEDs [26].

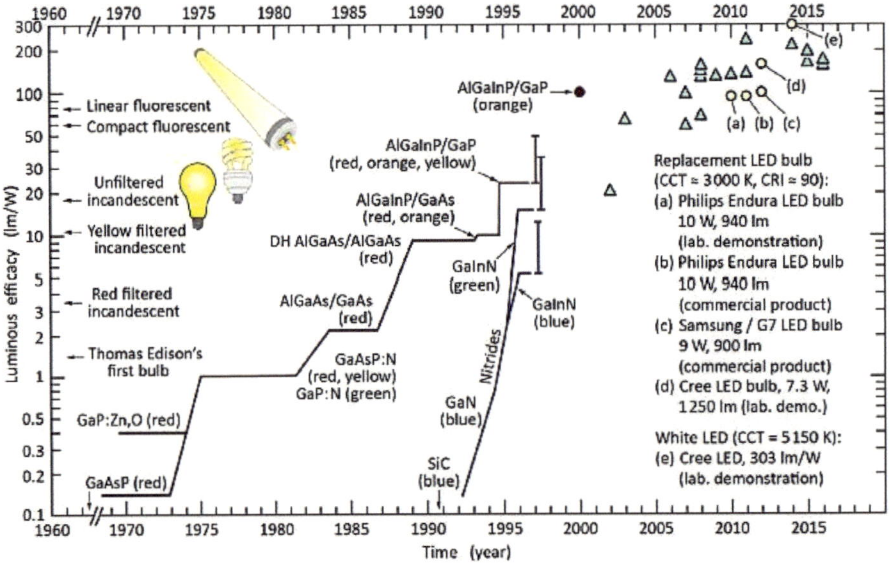

Figure 3.
The SSL LEDs efficiency improvement at laboratory demonstration and comparison with incandescent and fluorescent bulb efficiencies [4].

push blue GaN LEDs EQE to exceed 80%, while emissions wavelengths move to the red spectra range. In 1996, Nakamura *et al*. demonstrated the first short wavelength (417 nm) GaN LD. Since then, the 405 nm violet InGaN laser source has been commercially successful for high-definition video and multi-layer data storage. Using GaN-based blue LDs increases the data storage capabilities per disc in the blue-ray systems [28]. For their significant contribution in developing III-nitride-based LEDs, Shuji Nakamura, Hiroshi Amano, and Isamu Akasaki received the prestigious Noble Prize in physics in 2014 [29].

Today, the III-nitride-based LEDs have been proven to be highly efficient, have a long lifetime, and are environment friendly [4, 30]. The III-nitride-based white LEDs have high efficiency of 200 lm/W, and are commercially available for light applications [4]. The III-nitride semiconductors-based LEDs are used in almost every traffic signal, medical application, cell phone display, monitor, house/street, car, and so on [2, 30, 31]. Haitz *et al.* addressed that SSL LED lighting decreases electrical power consumption by more than 50% and is directly related to reducing CO_2 gas emissions by approximately 200 Megatons per year [32].

2. Polar-InGaN/GaN for blue and green LEDs

Acceleration of blue and green LEDs development from InGaN/GaN quantum well (QW) started after the successful invention of p-type GaN by low-energy electron irradiation in late 1989 by Amano *et al.* [24]. This invention led to high quantum efficiency and output power. However, InGaN/GaN growth along *c*-plane sapphire [0001] induces a polarization effect which causes quantum confined Stark effect (QCSE) and carrier loss (droop effect). The QCSE reduces the probability of radiation recombination by separating the hole and electron wave function on the QW. Especially, QCSE and polarization effects are highly pronounced with high indium (In) content incorporation in the InGaN epilayer [29].

The InGaN has the advantage of covering a wide spectral range from blue, green, and ultraviolet by tuning energy from 0.7 to 3.4 eV. Theoretically, when the indium composition is between 15 and 20% and 25–30%, the emission wavelengths belong to blue and green LEDs, respectively. However, to cover ultraviolet emission, the indium mole fractions should increase to a large extent, increasing threading dislocation density in the active layer. High threading dislocation density reduces internal quantum efficiency; therefore, InGaN material is not suitable for ultraviolet spectral emission.

Growth of InGaN/GaN with more than a 30% indium composition is difficult due to the different MOCVD growth temperature window [2, 33]. High-quality GaN requires a relatively high growth temperature (1050°C), whereas InGaN needs 800°C. Therefore, MOCVD needs high control on temperature fluctuations for better stability growth and to avoid inhomogeneous indium composition.

2.1 Polarization effect: Spontaneous polarization

Asymmetry of the wurtzite structure in (0001) sapphire direction (lattice constant c/a < 1.63 A, the electronegativity between Ga and N is vastly different) causes spontaneous polarization in the III-N material system [34].

AlN material has the largest spontaneous polarization compared to GaN and InN, as shown in **Figure 4**. Theoretical calculation of spontaneous polarization of ternary material can be expressed by [35]:

$$P^{Sp}_{InxGa1-xN} = -0.042\,x - 0.034(1-x) + 0.037\,x(1-x) \qquad (1)$$

$$P^{Sp}_{AlxGa1-xN} = 0.09\,x - 0.034(1-x) + 0.019\,x(1-x) \qquad (2)$$

2.2 Polarization effect: Piezoelectric polarization

Piezoelectric polarization effects are generated in the III-N materials system when the heterostructure undergoes huge lattice distortion from biaxial tensile/compressive

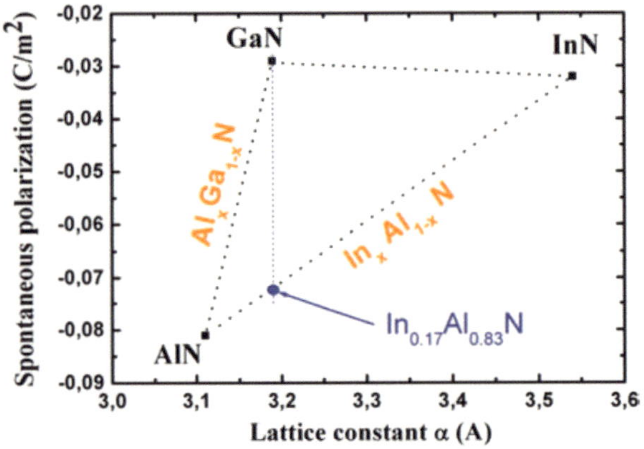

Figure 4.
Spontaneous polarization vs. lattice constant of III-N material system.

stress. The generated piezoelectric effect will be in [000–1] (tensile) / [0001] (compressive) direction [29, 36–39].

The piezoelectric effect from III-N ternary material can be calculated from biaxial stress and lattice mismatch. Therefore, piezoelectric polarization can be given as:

$$P_{PZ} = 2\left(e_{33}\varepsilon_z + e_{31}\left(\varepsilon_x + \varepsilon_y\right)\right) \tag{3}$$

where e_{33} and e_{31} are piezoelectric constants and ε_z is in-plane strain.

ε_z can be calculated from:

$$\varepsilon_z = \frac{\left(c - c_0\right)}{c_0} \tag{4}$$

where c, c_0 are in-plane lattice constant and relaxed lattice constant, respectively.

2.3 Quantum-confined stark effect

The polarization effect generates an electric field that eventually tilts the conduction and valance bands of InGaN/GaN multi-quantum wells (MQWs). Also, the electric field reduces spatial overlap between holes and electrons waves, thereby reducing the probability of radiative recombination, as shown in **Figure 5** [40, 41]. This effect is called QCSE, which highly depends on QW width. When the QW width increases, the QW luminance is red-shifted, whereas it is blue-shifted when the QW width decreases [42–46]. So, the effect of QCSE by polarization should consider two facts on efficient emission: the transition level and luminous intensity of InGaN QW.

First, the impact of transition level is a function of several parameters, such as QW width, biaxial stress, growth temperature, sub-band energy levels, etc. Since the spontaneous polarization constant difference is more negligible for GaN and InN, and lattice mismatch is significant, the contribution of piezoelectric polarization is dominant in InGaN ternary alloy. As the lattice mismatch increases, the electron and hole transition level decreases. Secondly, the electroluminescence intensity of InGaN/GaN MQW highly depends on the QW width. Electrons and holes are spatially separated when the QW width increases, reducing the wave function overlap. So, the peak wavelengths have blue-shift and eventually reduce luminescence intensity.

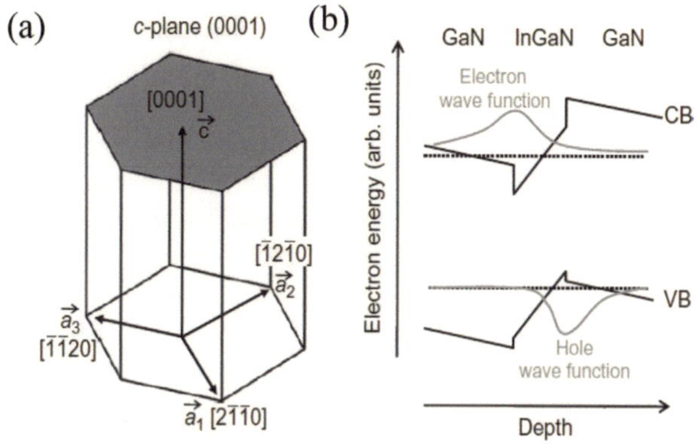

Figure 5.
(a) Wurtzite crystal structure of GaN (c-plane), and (b) electron and holes wavefunction separation due to polarization effect.

Due to lattice mismatch, huge screw and threading dislocations and point defects will be present in the QW layer. These defects can cause fluctuations that confine carriers in the minimum value of potential energy. Three significant facts potentially vary the localization of carriers and luminescence property of InGaN/GaN LEDs: indium clustering, well width fluctuations, and ternary alloy variations [46–48].

2.4 Methods to improve green LEDs efficiency using polar surface

Epitaxial growth of InGaN/GaN MQW structures on the polar substrate can reach more than 80% internal quantum efficiency (IQE). However, as the wavelength increases, InGaN/GaN-based LEDs' efficiency decreases by around 30%. Also, the efficiency of green LEDs decreases faster as the current injection increases. This problem leads to improper light mixing efficiency for white LEDs. Therefore, increasing the

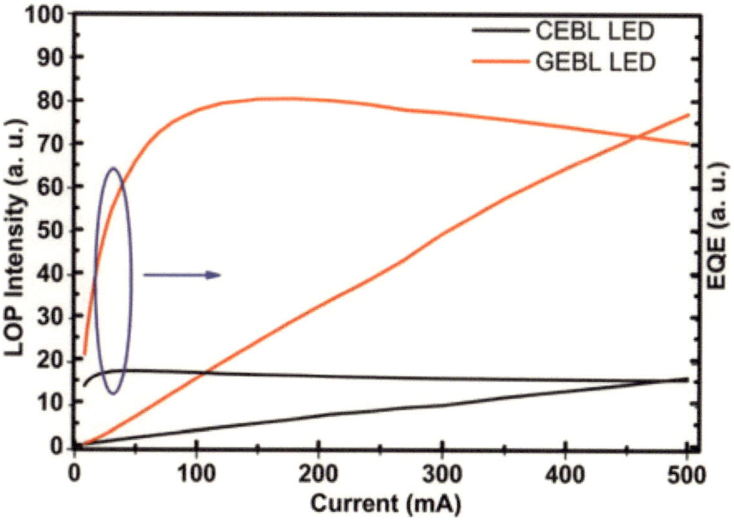

Figure 6.
Optical output power and EQE vs. current densities.

efficiency of green LEDs is the primary concern of RGB-LEDs. However, using the existing epitaxial growth technique, few methods can improve the green-based LED IQE, **Figure 6** continuous improvisation of epitaxial materials quality with less threading dislocation in the QW region, inserting pre-strain layer to accommodate lattice strain, and sharp QW interface by growth interruption method [29]. On the other hand, growing green LEDs epilayers on non-polar and semipolar substrates reduce the piezoelectric effect at the QW and avoid blue-shifted wave emission. Moreover, it increases the probability of radiative recombination to attain maximum IQE [49, 50].

3. Recent advancements in non-polar and Semipolar LEDs

Many researchers are developing the technology to reduce the piezoelectric polarization electric field effect in the active region of MQW devices. However, few methods have been proposed to increase the quantum efficiency of green LEDs to 163% by introducing a gradient electron blocking layer.

3.1 Non-polar LEDs

Several non-polar crystal planes occur in a wurtzite structure orthogonal to the basal planes. The non-polar *m*-planes $\{10\bar{1}0\}$ and *a*-planes $\{11\bar{2}0\}$ have zero inherent spontaneous polarization effect. InGaN QW grown on non-polar GaN orientation exhibits flat energy bands due to zero strain-induced piezoelectric effect, as shown in **Figure 7**. The device grown on an *m*-plane displays high brightness and one hundred times less decay [49, 50].

Researchers at University of California Santa Barbara (UCSB) established violet LEDs based on *m*-plane, which has low defect density, and the device performance is comparable with *c*-plane devices [51]. In addition, more uniform current injection, low current droop, and small blue-shift were observed for *m*-plane-oriented devices [52]. On the other hand, due to the lack of QCSE of non-polar InGaN/GaN MQW, the QW width can be increased without a significant loss in radiative recombination [52, 53].

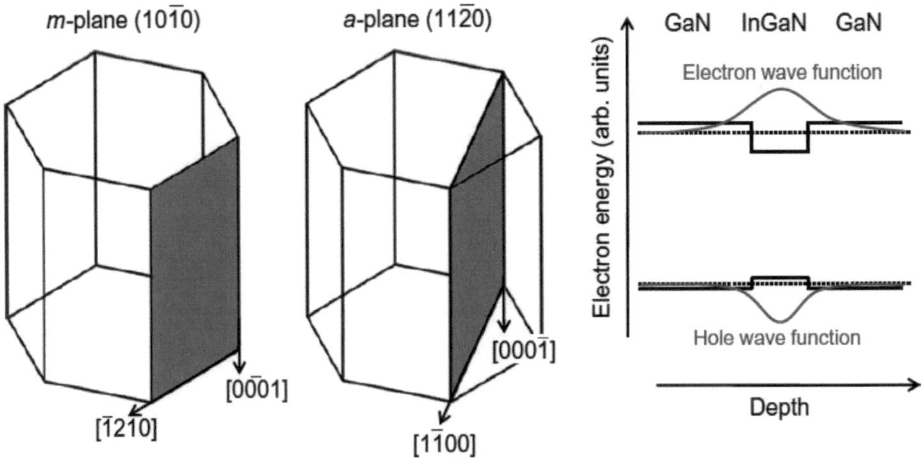

Figure 7.
The m-plane and a-plane of GaN wurtzite crystal structure. Wave function overlaps due to a lack of polarization for better radiative recombination.

However, the growth window for non-polar GaN is narrow and indium incorporation is limited with non-polar InGaN material [49].

Experimentally, it is well proven that growth of m-plane and a-plane GaN on non-polar direction reduces the spontaneous polarization effect. These planes are widely grown on sapphire or silicon substrates which normally shows five times higher (0.5 ~ 0.6 which is equivalent to $10^{10}/cm^2$) full width half maximum (FWHM) compared to that of grown on c-sapphire substrate. This high FWHM indicates a very high treading dislocation and stacking fault in GaN layer which reduces the crystalline and optical quality of GaN epilayer. Generally, the growth rate in a- or m- direction is lower when compared to c-plane direction. Due to this uneven growth rate the striation is seen in GaN epilayer. During the growth of non-polar GaN, it is important to supress the growth rate in vertical c-plane direction to have smooth morphology by avoiding island formation. The separation of c-plane and enhance of a-plane growth could be achieved in narrow growth temperature window [54–56].

3.2 Semipolar LEDs

The semipolar plane has a growth plane between the c-plane and non-polar plane, as shown in **Figure 8**. The semipolar plane shows a less polarization electric field effect than the c-plane [39, 57]. Semipolar directional growth has the advantage of high indium incorporation in the InGaN epi layer, which can be used for longer wavelengths. The total polarization and polarization discontinuity between InGaN and GaN will be dependent on the angle between the c-plane and the InGaN epilayer.

The motivation of using semipolar plane with different arbitrary angles is to reduce the polarization electric field to suppress the QCSE effect, as shown in **Figure 9**. Among these semipolar planes, $\{\bar{2}021\}$ orientation showed better performance due to high compositional homogeneity across QW [58]. The semipolar directional growth yields a low polarization electric field; however, stacking fault formation is inevitable due to the material's anisotropy property. The first semipolar LED was demonstrated on $\{30\bar{3}8\}$ while the first semipolar green LED was demonstrated on GaN along $\{10\bar{1}3\}$ direction [59, 60]. Currently, the IQE of c-plane-based yellow-green LEDs is surpassed by semipolar

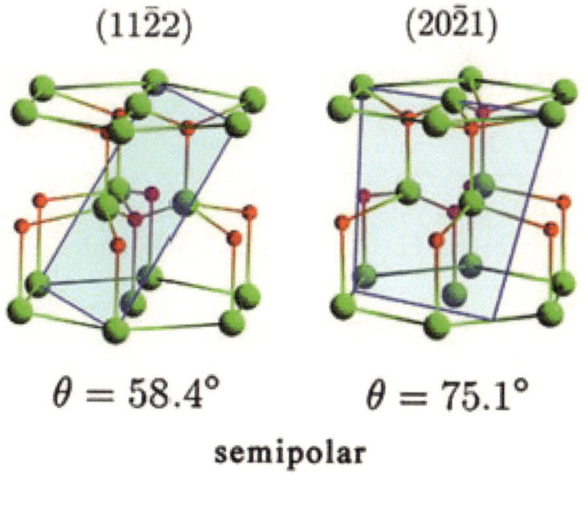

Figure 8.
Two different Semipolar wurtzite structures of GaN [49].

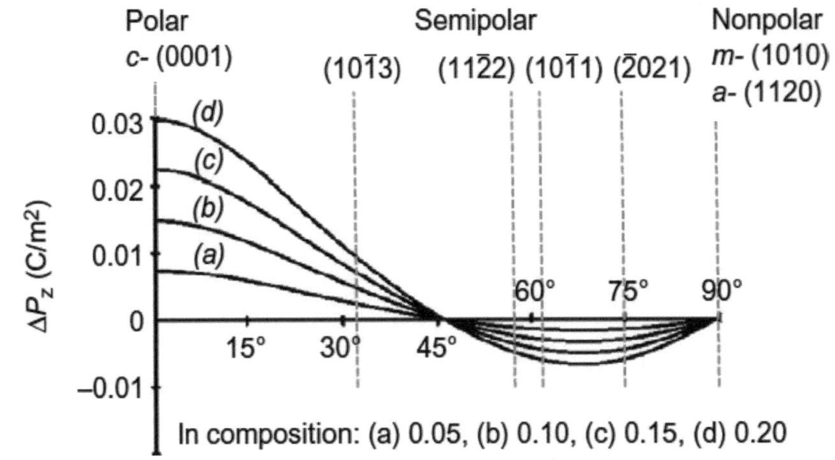

Figure 9.
Polarization discontinuity of GaN for various arbitrary angles with respect to the c-plane [50].

Non-polar	Semi-polar
Only two different plane orientations	Vast variants of planes
Gives maximum wave function overlap	Substantial wave function overlap (comparatively high than *c*-plane)
Absence of QCSE	QCSE effect presents except few planes (e.g $20\bar{2}\bar{1}$)
Indium is difficult to incorporate which reduces the growth window	Indium incorporation is high which enhances the growth window
Not suitable for long wavelength emission due to defects in QWs	Promising for longer wavelength emission
Larger degree of polarized emission	Degree of polarization is comparatively less than non-polar

Table 1.
The important properties of non-polar and semi-polar GaN LEDs devices.

long wavelengths of yellow-green LEDs. In 2011, Mg-doped quantum barriers homoepitaxially grown on GaN increased radiative recombination probability to improve green LEDs' performance [61]. Researchers at UCSB have reported output power of 31.1 mW and 9.9 mW at 20 mA, with EQE of 54% and 20.4% of blue and green LEDs, respectively. These blue and green LEDs were grown on a semipolar growth plane of $\{10\bar{1}1\}$ and $\{11\bar{2}2\}$, respectively [62, 63]. The significant differences between non-polar and semi-polar based GaN LEDs devices are given in **Table 1**.

4. Micro LED

Display-based LED technology is considered the most promising for next-generation display devices. Every display screen consists of several small fractions known as a pixel, and the size of these pixels determines the resolution of the display screen. The resolution of the display screen is measured in the term of PPI, which means pixel per square inch. We could estimate this PPI term by comparing the resolution of conventional Organic Light Emitting Diode (OLED) displays and modern smartphones; commercial OLED televisions currently have a pixel density of about 100 to 200 PPI, while the resolution of new smartphones is around 400 to 500 PPI. This pixel density is proportional to the size of a single light emitting unit, which in this case is in the sub-millimeter range. The resolution of the micro-LED display can be increased by fabricating sub-micron LEDs. The individual LEDs would be one pixel for the display, and the size of the LEDs could be reduced according to the desired resolution. In addition, these micro-LEDs have fast response, and are energy efficient than previously developed display technology. The OLED technology, which was developed in the early 90s, was ahead of its time, and during the later decade, it was commercially available in some consumer displays. Its significant advantages over the LCD were self-luminous, wide viewing angle, high contrast, and fast response.

On the other hand, these displays were not used widely in consumer electronics [64–66]. This event proved to be a fracture in display technology; major market players and researchers started looking for other materials and technology. However, the main objective remains the same: high contrast, efficiency, resolution, and the process should be well established. Currently, the research is focused on developing micro-LEDs of size <100 μm for high-resolution. Furthermore, according to market research, it is estimated that the micro-LED display market will grow by around 20.5 billion USD in 2025 with an annual growth rate of about 80%.

Jiang's group did the first endeavor for the growth of micro-LED. During their quest, they could fabricate a blue GaN micro-LED chip with an individual pixel diameter of 12 μm [67]. The display made from these pixels measured 0.5 X 0.5 mm^2 and had 10 × 10 pixels. Soon after, micro-LED became a hot topic among researchers, leading to the further development of the fabrication technology. Thenceforth, the researchers have extensively understood the nature and epitaxy of III-N material. This has resulted in efficient and complex structures of GaN-based LEDs. Modern LED structures may consist of many layers up to 100 and a lifetime of more than fifty thousand hours. In 2014, Christian et al. fabricated a 50 X 50 μm^2 LED and successfully transferred it to a flexible substrate. The fabricated LED emits 60 μW optical power at 1 mA current injection [68]. This research paves the path towards a flexible screen, which overcomes the limitation that III-N micro-LED faced while competing with OLED technology.

The color contrast of the micro-LED display is based on three primary colors (red, green, and blue, also called RGB) that can be combined in different ratios to give all colors in nature. Thus, combining three micro-LED (RGB) on a single pixel solves the problem. **Figure 10** shows the schematic of such pixel consisting of RGB micro-LEDs. In this arrangement, the different currents would be applied to control the brightness of each LED to realize the combination of RGB primary color to achieve the full-color display. However, integrating three different material LEDs on the same pixel is difficult to mass produce. For example, if there is a need to fabricate a 4 K resolution display, nearly 25 million micro-LEDs are needed to assemble or fabricate. This

Figure 10.
Schematic of RGB micro-LED full-color pixel.

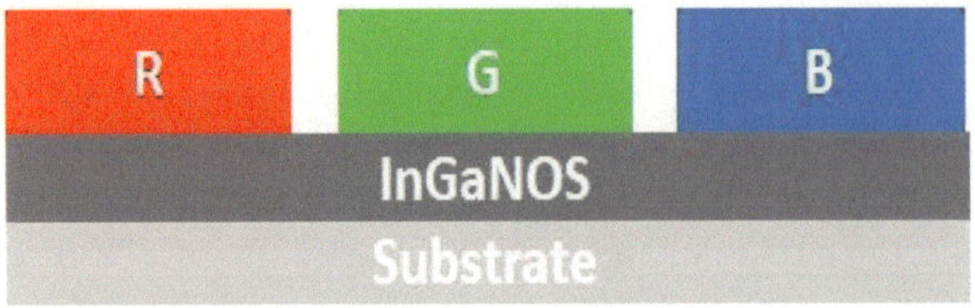

Figure 11.
Schematic of InGaN LED structures grown on InGaNOS modified substrate.

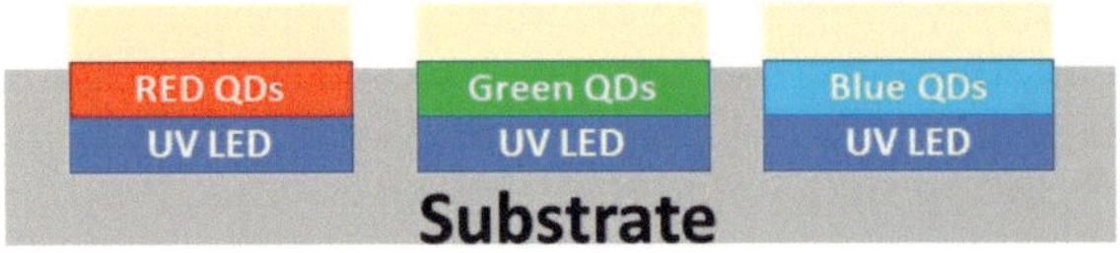

Figure 12.
Schematic of UV micro-LED and QDs integration approach to realize the full-color pixel.

hurdle could be overcome by fabricating all LEDs of the same material. The solution is fabricating the GaN/InGaN base LEDs where the emission wavelength could be tuned by varying the molar ration of In in InGaN alloy. The said approach was developed by A. Even et al. used a very innovative substrate to grow InGaN base LEDs. The unique substrate is InGaNOS (InGaN on pseudo-substrate), which overcomes the lattice mismatch and reduces the stress in as-grown films. The epitaxy results show that the grown LED structures could cover blue (482 nm) to red (617 nm) [69]. The schematic is shown in **Figure 11**. Similarly, Pasayat et al. fabricated an LED structure on GaN porous pseudo substrate. The fabricated LEDs range from 8 x 8 μm^2 to 20 x 20 μm^2. It was observed that the reduction in size leads to an EL redshift from 525 to 561 nm [70]. As per the study, the color-tunable monolithic integration nitride-based RGB technique can fabricate micro-LED arrays.

Another efficient approach was developed by Prof. Kuo, in which the full-color display pixel could be achieved by employing UV LEDs and a down-conversion QDs (Quantum Dots) solution. In this approach, all three pixels are fabricated as AlGaN/GaN-based UV LED, and QDs are placed above them. The targeted QDs down convert UV emission to respective wavelength, i.e., RGB. The schematic of this approach is shown in **Figure 12** [71, 72].

5. Conclusion

We reviewed the history of GaN epitaxy, III-N material improvement strategies, and the advancement of LEDs. It is necessary to reduce the MQW threading dislocation density, QCSE effect, and carrier localization effect to enhance the IQE. To get a high hole concentration by manipulating the Fermi level, we must design strain compensated InGaN/AlInGaN and doped InGaN/GaN barriers. For *c*-plane growth, ultra-sharp (thickness ~ 3 nm) QW structure with low defect density can yield high IQE. However, for good wavefunction matching and low blue and redshift on the spectral emission, the non-polar and semipolar substrate can be a potential candidate for the reported output power of 31.1 mW and 9.9 mW at 20 mA, with EQE of 54% and 20.4% of blue and green LEDs, respectively. For non-polar GaN epitaxy of $\{11\bar{2}0\}$ *a*-plane and $\{10\bar{1}0\}$ *m*-planes can be grown on *r*-plane Al_2O_3, *a*-plane SiC, and *m*-plane SiC, (100), $LiAlO_2$ are the promising epitaxial substrates. For Semi-polar GaN epitaxy, $\{\bar{2}021\}$ orientation on patterned *r*-plane Al_2O_3 substrates showed better performance due to their high compositional homogeneity across QW. With the recent developments, GaN is not limited to lighting, but the latest innovation also led to the development of micro-LEDs, laser projection, and point sources. These developments have pushed GaN into the realm of display technology. The miniaturization of the GaN-based micro-LEDs and integration of GaN on silicon will help drive the application into photonic integrated circuits (ICs).

Author details

Thamer A. Tabbakh[1*], Deepak Anandan[2], Michael J. Sheldon[3], Prashant Tyagi[2] and Ahmad Alfaifi[4]

1 National Center for Nanotechnology and Semiconductor, KACST, Riyadh, Saudi Arabia

2 Skyline Semiconductor Services, Riyadh, Saudi Arabia

3 Orbit Engineering, California, USA

4 Saudi Semiconductors Programm, KACST, Riyadh, Saudi Arabia

*Address all correspondence to: t.tabbakh@gmail.com

References

[1] Seong T-Y, Han J, Amano H, Morkoc H, editors. III-Nitride Based Light Emitting Diodes and Applications. Vol. 126. Dordrecht: Springer Netherlands; 2013. DOI: 10.1007/978-94-007-5863-6

[2] Ambacher O. Growth and applications of group III-nitrides. Journal of Physics D: Applied Physics. 1998;**31**(20):2653-2710. DOI: 10.1088/0022-3727/31/20/001

[3] Strite S. GaN, AlN, and InN: A review. Journal of Vacuum Science & Technology B: Microelectronics and Nanometer Structures. 1992;**10**(4):1237. DOI: 10.1116/1.585897

[4] Cho J, Park JH, Kim JK, Schubert EF. White light-emitting diodes: History, progress, and future: White light-emitting diodes. Laser & Photonics Reviews. 2017;**11**(2):1600147. DOI: 10.1002/lpor.201600147

[5] Chen Y et al. GaN in different dimensionalities: Properties, synthesis, and applications. Materials Science & Engineering R: Reports. 2019;**138**:60-84. DOI: 10.1016/j.mser.2019.04.001

[6] Bass M, Optical Society of America, editors. Handbook of Optics. 2nd ed. New York: McGraw-Hill; 1995

[7] Yoshida H, Kuwabara M, Yamashita Y, Uchiyama K, Kan H. The current status of ultraviolet laser diodes: The current status of ultraviolet laser diodes. Physica Status Solidi A: Applications and Materials Science. 2011;**208**(7):1586-1589. DOI: 10.1002/pssa.201000870

[8] Amano H et al. The 2018 GaN power electronics roadmap. Journal of Physics D: Applied Physics. 2018;**51**(16):163001. DOI: 10.1088/1361-6463/aaaf9d

[9] Utsumi W, Saitoh H, Kaneko H, Watanuki T, Aoki K, Shimomura O. Congruent melting of gallium nitride at 6 GPa and its application to single-crystal growth. Nature Materials. 2003;**2**(11):735-738. DOI: 10.1038/nmat1003

[10] Porowski S, Grzegory I. Thermodynamical properties of III–V nitrides and crystal growth of GaN at high N2 pressure. Journal of Crystal Growth. 1997;**178**(1-2):174-188. DOI: 10.1016/S0022-0248(97)00072-9

[11] Karpiński J, Jun J, Porowski S. Equilibrium pressure of N2 over GaN and high pressure solution growth of GaN. Journal of Crystal Growth. 1984;**66**(1):1-10. DOI: 10.1016/0022-0248(84)90070-8

[12] Grandjean N, Massies J, Leroux M. Nitridation of sapphire. Effect on the optical properties of GaN epitaxial overlayers. Applied Physics Letters. 1996;**69**(14):2071-2073. DOI: 10.1063/1.116883

[13] Heinlein C, Grepstad J, Berge T, Riechert H. Preconditioning of c-plane sapphire for GaN epitaxy by radio frequency plasma nitridation. Applied Physics Letters. 1997;**71**(3):341-343. DOI: 10.1063/1.119532

[14] Dixit R et al. Influence of growth temperature on laser molecular beam epitaxy and properties of GaN layers grown on c-plane sapphire. Optical Materials. 2017;**66**:142-148. DOI: 10.1016/j.optmat.2017.01.053

[15] Tyagi P et al. Dependence of Al incorporation on growth temperature during laser molecular beam epitaxy of AlxGa1−xN epitaxial layers on sapphire (0001). Journal of Alloys

and Compounds. 2018;**739**:122-128. DOI: 10.1016/j.jallcom.2017.12.220

[16] Johnson WC, Parson JB, Crew MC. Nitrogen compounds of gallium. III. The Journal of Physical Chemistry. 1932;**36**(10):2651-2654. DOI: 10.1021/j150340a015

[17] Maruska HP, Tietjen JJ. THE PREPARATION AND PROPERTIES OF VAPOR-DEPOSITED SINGLE-CRYSTAL-LINE GaN. Applied Physics Letters. 1969;**15**(10):327-329. DOI: 10.1063/1.1652845

[18] Manasevit HM, Simpson WI. The use of metal-organics in the preparation of semiconductor materials. Journal of the Electrochemical Society. 1969;**116**(12):1725. DOI: 10.1149/1.2411685

[19] Kim W, Botchkarev AE, Salvador A, Popovici G, Tang H, Morkoç H. On the incorporation of Mg and the role of oxygen, silicon, and hydrogen in GaN prepared by reactive molecular beam epitaxy. Journal of Applied Physics. 1997;**82**(1):219-226. DOI: 10.1063/1.365801

[20] Seifert W, Franzheld R, Butter E, Sobotta H, Riede V. On the origin of free carriers in high-conducting n-GaN. Crystal Research and Technology. 1983;**18**(3):383-390. DOI: 10.1002/crat.2170180314

[21] Yoshida S, Misawa S, Gonda S. Improvements on the electrical and luminescent properties of reactive molecular beam epitaxially grown GaN films by using AlN-coated sapphire substrates. Applied Physics Letters. 1983;**42**(5):427-429. DOI: 10.1063/1.93952

[22] Amano H, Sawaki N, Akasaki I, Toyoda Y. Metalorganic vapor phase epitaxial growth of a high quality GaN film using an AlN buffer layer. Applied Physics Letters. 1986;**48**(5):353-355. DOI: 10.1063/1.96549

[23] Nakamura S. GaN growth using GaN buffer layer. Japanese Journal of Applied Physics. 1991;**30**(Part 2, 10A):L1705-L1707. DOI: 10.1143/JJAP.30.L1705

[24] Amano H, Kito M, Hiramatsu K, Akasaki I. P-type conduction in Mg-doped GaN treated with low-energy electron beam irradiation (LEEBI). Japanese Journal of Applied Physics. 1989;**28**(Part 2, 12):L2112-L2114. DOI: 10.1143/JJAP.28.L2112

[25] Nakamura S, Mukai T, Senoh M, Iwasa N. Thermal annealing effects on P-type Mg-doped GaN films. Japanese Journal of Applied Physics. 1992;**31**(Part 2, 2B):L139-L142. DOI: 10.1143/JJAP.31.L139

[26] Nakamura S, Krames MR. History of gallium–nitride-based light-emitting diodes for illumination. Proceedings of the IEEE. 2013;**101**(10):2211-2220. DOI: 10.1109/JPROC.2013.2274929

[27] Nakamura S, Mukai T. High-quality InGaN films grown on GaN films. Japanese Journal of Applied Physics. 1992;**31**(Part 2, 10B):L1457-L1459. DOI: 10.1143/JJAP.31.L1457

[28] Nam OH et al. Characteristics of GaN-based laser diodes for post-DVD applications. Physica Status Solidi (a). 2004;**201**(12):2717-2720. DOI: 10.1002/pssa.200405114

[29] Wallace M. Optoelectronic Study of InGaN/GaN LEDs. [Thesis]. University of Strathclyde; 2016. DOI: 10.48730/43Q0-ZY87. Available from: https://stax.strath.ac.uk/concern/theses/nv935295j

[30] Amano H. Nobel lecture: Growth of GaN on sapphire via low-temperature

deposited buffer layer and realization of p -type GaN by Mg doping followed by low-energy electron beam irradiation. Reviews of Modern Physics. 2015;**87**(4):1133-1138. DOI: 10.1103/RevModPhys.87.1133

[31] Shimada J, Kawakami Y, Fujita S. Development of lighting goggles with power white LED modules. In: Proceedings of SPIE - SPIE Digital Library. San Jose, CA. 2003. p. 174. DOI: 10.1117/12.476558

[32] Haitz R, Kish F, Tsao J, Nelson J. The Case for a National Research Program on Semiconductor Lighting. Sandia Report SAND-2000 1612. 2000. Available from: https://www.sandia.gov/app/uploads/sites/153/2021/12/hpsnl_sand_report_2000.pdf

[33] Dupuis RD. Epitaxial growth of III–V nitride semiconductors by metalorganic chemical vapor deposition. Journal of Crystal Growth. 1997;**178**(1-2):56-73. DOI: 10.1016/S0022-0248(97)00079-1

[34] Bernardini F, Fiorentini V, Vanderbilt D. Spontaneous polarization and piezoelectric constants of III-V nitrides. Physical Review B. 1997;**56**(16):R10024-R10027. DOI: 10.1103/PhysRevB.56.R10024

[35] Bernardini F, Fiorentini V. Nonlinear behavior of spontaneous and piezoelectric polarization in III-V nitride alloys. Physica Status Solidi A: Applications and Materials Science. 2002;**190**(1):65-73. DOI: 10.1002/1521-396X(200203)190:1<65::AID-PSSA65>3.0.CO;2-0

[36] Doshi B, Brennan KF, Bicknell-Tassius R, Grunthaner F. The effect of strain-induced polarization fields on impact ionization in a multiquantum-well structure. Applied Physics Letters. 1998;**73**(19):2784-2786. DOI: 10.1063/1.122590

[37] Deng Y et al. Study on strain and piezoelectric polarization of AlN thin films grown on Si. Journal of Vacuum Science & Technology A: Vacuum, Surfaces, and Films. 2005;**23**(4):628-630. DOI: 10.1116/1.1927533

[38] Romanov AE, Baker TJ, Nakamura S, Speck JS, ERATO/JST UCSB Group. Strain-induced polarization in wurtzite III-nitride semipolar layers. Journal of Applied Physics. 2006;**100**(2):023522. DOI: 10.1063/1.2218385

[39] Gil B, Bigenwald P, Briot O. Polarization fields in wurtzite strained layers grown on () planes. Superlattices and Microstructures. 2008;**44**(3):291-301. DOI: 10.1016/j.spmi.2008.06.003

[40] Della Sala F et al. Free-carrier screening of polarization fields in wurtzite GaN/InGaN laser structures. Applied Physics Letters. 1999;**74**(14):2002-2004. DOI: 10.1063/1.123727

[41] Seo Im J, Kollmer H, Off J, Sohmer A, Scholz F, Hangleiter A. Reduction of oscillator strength due to piezoelectric fields in G a N/a l x Ga 1 − x N quantum wells. Physical Review B. 1998;**57**(16):R9435-R9438. DOI: 10.1103/PhysRevB.57.R9435

[42] Peter M, Laubsch A, Stauss P, Walter A, Baur J, Hahn B. Green ThinGaN power-LED demonstrates 100 lm. Physica Status Solidi C: Current Topics in Solid State Physics. 2008;**5**(6):2050-2052. DOI: 10.1002/pssc.200778554

[43] Lutgen S et al. True green InGaN laser diodes. Physica Status Solidi A: Applications and Materials Science. 2010;**207**(6):1318-1322. DOI: 10.1002/pssa.200983620

[44] Queren D et al. 500 nm electrically driven InGaN based laser diodes. Applied

Physics Letters. 2009;**94**(8):081119. DOI: 10.1063/1.3089573

[45] Queren D et al. Quality and thermal stability of thin InGaN films. Journal of Crystal Growth. 2009;**311**(10):2933-2936. DOI: 10.1016/j.jcrysgro.2009.01.066

[46] Brillson LJ, Levin TM, Jessen GH, Ponce FA. Localized states at InGaN/GaN quantum well interfaces. Applied Physics Letters. 1999;**75**(24):3835-3837. DOI: 10.1063/1.125472

[47] Golub LE et al. Low-temperature kinetics of localized excitons in quantum-well structures. Physica Status Solidi B: Basic Solid State Physics. 1998;**205**(1):203-208. DOI: 10.1002/(SICI)1521-3951(199801)205:1<203:AID-PSSB203>3.0.CO;2-N

[48] Na JH et al. Dependence of carrier localization in InGaN/GaN multiple-quantum wells on well thickness. Applied Physics Letters. 2006;**89**(25):253120. DOI: 10.1063/1.2423232

[49] Li J et al. III-Nitrides Light Emitting Diodes: Technology and Applications. Vol. 306. Singapore: Springer Singapore; 2020. DOI: 10.1007/978-981-15-7949-3

[50] Kelchner KM, DenBaars SP, Speck JS. GaN laser diodes on nonpolar and Semipolar planes. In: Semiconductors and Semimetals. Vol. 86. Elsevier; 2012. pp. 149-182. DOI: 10.1016/B978-0-12-391066-0.00004-6

[51] Yamada H et al. Comparison of InGaN/GaN light emitting diodes grown on m -plane and a -plane bulk GaN substrates: Comparison of InGaN/GaN light emitting diodes grown on m -plane and a -plane bulk GaN substrates. Physics Status Solidi RRL - Rapid Research Letter. 2008;**2**(2):89-91. DOI: 10.1002/pssr.200701313

[52] Ling S-C, Lu T-C, Chang S-P, Chen J-R, Kuo H-C, Wang S-C. Low efficiency droop in blue-green m-plane InGaN/GaN light emitting diodes. Applied Physics Letters. 2010;**96**(23):231101. DOI: 10.1063/1.3449557

[53] Lin Y-D et al. Characterization of blue-green m-plane InGaN light emitting diodes. Applied Physics Letters. 2009;**94**(26):261108. DOI: 10.1063/1.3167824

[54] Cai Y, Shen S, Zhu C, Zhao X, Bai J, Wang T. Nonpolar ($11\bar{2}0$) GaN metal–semiconductor–metal photodetectors with superior performance on silicon. ACS Applied Materials & Interfaces. 2020;**12**(22):25031-25036. DOI: 10.1021/acsami.0c04890

[55] Izyumskaya N et al. Epitaxial lateral overgrowth of non-polar GaN($1\bar{1}00$) on Si(112) patterned substrates by MOCVD. Journal of Crystal Growth. 2011;**314**(1):129-135. DOI: 10.1016/j.jcrysgro.2010.11.152

[56] Jiu L, Gong Y, Wang T. Overgrowth and strain investigation of (11-20) non-polar GaN on patterned templates on sapphire. Scientific Reports. 2018;**8**(1):9898. DOI: 10.1038/s41598-018-28328-7

[57] Takeuchi T, Amano H, Akasaki I. Theoretical study of orientation dependence of piezoelectric effects in Wurtzite strained GaInN/GaN Heterostructures and quantum Wells. Japanese Journal of Applied Physics. 2000;**39**(Part 1, 2A):413-416. DOI: 10.1143/JJAP.39.413

[58] Ueno M et al. InGaN-based true green laser diodes on novel semi-polar GaN substrates. Journal of Crystal Growth. 2011;**315**(1):258-262. DOI: 10.1016/j.jcrysgro.2010.07.016

[59] Kamiyama S et al. GaN growth on (30&3macr,8) 4H-SiC substrate for reduction of internal polarization. Physica Status Solidi C: Current Topics in Solid State Physics. 2005;**2**(7):2121-2124. DOI: 10.1002/pssc.200461390

[60] Sharma R et al. Demonstration of a semipolar (10¯1¯3¯) InGaN/GaN green light emitting diode. Applied Physics Letters. 2005;**87**(23):231110. DOI: 10.1063/1.2139841

[61] Huang C-Y et al. Influence of Mg-doped barriers on semipolar (202¯1) multiple-quantum-well green light-emitting diodes. Applied Physics Letters. 2011;**99**(14):141114. DOI: 10.1063/1.3647560

[62] Zhao Y et al. 30-mW-class high-power and high-efficiency blue Semipolar (10\bar1\bar1) InGaN/GaN light-emitting diodes obtained by backside roughening technique. Applied Physics Express. 2010;**3**(10):102101. DOI: 10.1143/APEX.3.102101

[63] Yamamoto S et al. High-efficiency single-quantum-well green and yellow-green light-emitting diodes on Semipolar (20\bar21) GaN substrates. Applied Physics Express. 2010;**3**(12):122102. DOI: 10.1143/APEX.3.122102

[64] Geffroy B, le Roy P, Prat C. Organic light-emitting diode (OLED) technology: Materials, devices and display technologies. Polymer International. 2006;**55**(6):572-582. DOI: 10.1002/pi.1974

[65] Chen H, Tan G, Wu S-T. Ambient contrast ratio of LCDs and OLED displays. Optics Express. 2017;**25**(26):33643. DOI: 10.1364/OE.25.033643

[66] Wu T et al. Mini-LED and micro-LED: Promising candidates for the next generation display technology. Applied Sciences. 2018;**8**(9):1557. DOI: 10.3390/app8091557

[67] Jiang HX, Jin SX, Li J, Shakya J, Lin JY. III-nitride blue microdisplays. Applied Physics Letters. 2001;**78**(9):1303-1305. DOI: 10.1063/1.1351521

[68] Goßler C et al. GaN-based micro-LED arrays on flexible substrates for optical cochlear implants. Journal of Physics D: Applied Physics. 2014;**47**(20):205401. DOI: 10.1088/0022-3727/47/20/205401

[69] Even A et al. Enhanced In incorporation in full InGaN heterostructure grown on relaxed InGaN pseudo-substrate. Applied Physics Letters. 2017;**110**(26):262103. DOI: 10.1063/1.4989998

[70] Pasayat SS et al. Color-tunable < 10μm square InGaN micro-LEDs on compliant GaN-on-porous-GaN pseudo-substrates. Applied Physics Letters. 2020;**117**(6):061105. DOI: 10.1063/5.0011203

[71] Han H-V et al. Resonant-enhanced full-color emission of quantum-dot-based micro LED display technology. Optics Express. 2015;**23**(25):32504. DOI: 10.1364/OE.23.032504

[72] Lin H-Y et al. Optical cross-talk reduction in a quantum-dot-based full-color micro-light-emitting-diode display by a lithographic-fabricated photoresist mold. Photonics Research. 2017;**5**(5):411. DOI: 10.1364/PRJ.5.000411